环境工程专业英语

主 编 赵锐 刘洋

西南交通大学出版社
·成都·

图书在版编目（CIP）数据

环境工程专业英语 / 赵锐，刘洋主编. —成都：西南交通大学出版社，2022.6（2024.1 重印）
ISBN 978-7-5643-8712-9

Ⅰ. ①环… Ⅱ. ①赵… ②刘… Ⅲ. ①环境工程 – 英语 – 高等学校 – 教材 Ⅳ. ①X5

中国版本图书馆 CIP 数据核字（2022）第 093366 号

Huanjing Gongcheng Zhuanye Yingyu
环境工程专业英语

主　编／赵锐　刘洋　　　　　　责任编辑／牛　君
　　　　　　　　　　　　　　　　封面设计／原谋书装

西南交通大学出版社出版发行
（四川省成都市金牛区二环路北一段 111 号西南交通大学创新大厦 21 楼　610031）
营销部电话：028-87600564　　028-87600533
网址：http://www.xnjdcbs.com
印刷：成都蜀雅印务有限公司

成品尺寸　185 mm × 260 mm
印张　14.75　　字数　455 千
版次　2022 年 6 月第 1 版　　印次　2024 年 1 月第 2 次

书号　ISBN 978-7-5643-8712-9
定价　46.00 元

课件咨询电话：028-81435775
图书如有印装质量问题　本社负责退换
版权所有　盗版必究　举报电话：028-87600562

前言

近年来我国高等工程教育不断发展，工程教育规模位居世界首位。特别是2016年，在国际工程联盟大会上，我国成为《华盛顿协议》的第18个正式成员，这对我国高等工程教育教学质量提出了新的要求。培养具有家国情怀、国际视野、跨文化沟通能力、服务国家乃至全球环境工程领域的高端技术人才和工程管理人才已成为高等工程教育的内部需求。

环境工程是一门与土木建筑、化学工程、生物学、气象学、管理学和社会学等多门学科相关的交叉学科，它通过评价人类生产和社会活动对环境的影响，用具体的工程、规划和管理措施，控制环境污染，保护环境与资源，使社会、经济和环境协调发展。

环境工程专业英语教材的编写不仅要涵盖环境工程基本专业知识，还要兼顾学术英语阅读、翻译、写作以及学术交流等应用能力的培养，同时为学习者打开环境工程专业的国际视野。本书体例上以 Unit 为单位，共 6 个 Unit，内容涵盖了水环境、大气环境、土壤环境污染控制，环境影响评价，环境规划与管理以及可持续发展等环境工程专业知识。全书选材均出自近年出版的原版教材、专著、科技文献资料以及国际环境管理机构官方网站。每个 Unit 独立讲述某一方面的内容，方便教师根据需要灵活掌握授课重点。书中每一个 Unit 包括 Reading, Translation, Writing 等 3 部分。同时，课文后列出了文中出现的新词汇和短语的释义，并对课文中的疑难词汇或句子进行了讲解。各个部分都有习题供学习者进行练习，以巩固学习成果。本书还提供了部分习题答案以及学术论文写作样本和学术海报的制作。

本书可作为环境工程专业本科生和研究生的专业英语教材，也可供相关学科本科生、研究生、科研人员以及相关人员自学参考，还可用作培训教材。

本书由赵锐和刘洋主编，刘洋负责编写第1、2、3、4单元以及所有的阅读习题和写作部分，赵锐负责编写第5、6单元。感谢西南交通大学地球科学与环境工程学院对本书编写和出版的关心和支持。

囿于编者知识背景与见识，加之编写时间紧促，书中难免存在疏漏与不妥之处，恳请广大读者不吝赐教。

编 者

2022年1月

Contents

Unit 1 Introduction to Environmental Engineering

Lesson 1 What is Environmental Engineering? ·············1
    Part 1 Reading············1
    Part 2 Translation············5
    Part 3 Writing············9
Lesson 2 Global Environmental Concerns············14
    Part 1 Reading············14
    Part 2 Translation············21
    Part 3 Writing············25

Unit 2 Air Pollution and Control

Lesson 1 Air Pollution, Source and Characteristics············28
    Part 1 Reading············28
    Part 2 Translation············33
    Part 3 Writing············36
Lesson 2 Air Pollution Emission Control Devices for Stationary Sources············41
    Part 1 Reading············41
    Part 2 Translation············48
    Part 3 Writing············51

Unit 3 Water Pollution and Control

Lesson 1 Processing of Water Supply············58
    Part 1 Reading············58
    Part 2 Translation············65
    Part 3 Writing············67
Lesson 2 Conventional Wastewater Treatment Process············71
    Part 1 Reading············71
    Part 2 Translation············79
    Part 3 Writing············84
Lesson 3 Advanced Wastewater Treatment············88
    Part 1 Reading············88
    Part 2 Translation············94
    Part 3 Writing············98

Unit 4　Solid Waste and Disposal

Lesson 1　Solid Wastes and the Management of Solid Wastes ··················102
　　Part 1　Reading ··················102
　　Part 2　Translation ··················110
　　Part 3　Writing ··················115
Lesson 2　Hazardous Waste Management ··················119
　　Part 1　Reading ··················119
　　Part 2　Translation ··················124
　　Part 3　Writing ··················129
Lesson 3　Soil Pollution ··················135
　　Part 1　Reading ··················135
　　Part 2　Translation ··················141
　　Part 3　Writing ··················146

Unit 5　Environmental Management and Environmental Impact Assessment

Lesson 1　Administration on Environmental Management ··················156
　　Part 1　Reading ··················156
　　Part 2　Translation ··················161
　　Part 3　Writing ··················164
Lesson 2　Environmental Impact Assessment (EIA) ··················173
　　Part 1　Reading ··················173
　　Part 2　Translation ··················178
　　Part 3　Writing ··················184

Unit 6　Environmental Laws, Regulations and Sustainable Development

Lesson 1　Governance — Water on the move ··················194
　　Part 1　Reading ··················194
　　Part 2　Translation ··················201
　　Part 3　Writing ··················206
Lesson 2　Global Megatrends ··················213
　　Part 1　Reading ··················213
　　Part 2　Translation ··················217
　　Part 3　Writing ··················222

练习题参考答案 ··················228
参考文献 ··················229

Unit 1 Introduction to Environmental Engineering

Lesson 1 What is Environmental Engineering?

Part 1 Reading

What is Environmental Engineering?

Environmental engineering is the integration of sciences and engineering principles to improve the natural environment, to provide healthy water, air, and land for human habitation and for other organisms, and to clean up pollution sites. Environmental engineering can also be described as a branch of applied science and technology that addresses the issue of energy preservation, production asset and control of waste from human and animal activities. Furthermore, it is concerned with finding plausible solutions in the field of public health, such as waterborne diseases, implementing laws which promote adequate sanitation in urban, rural and recreational areas. It involves waste water management and air pollution control, recycling, waste disposal, radiation protection, industrial hygiene, environmental sustainability, and public health issues as well as knowledge of environmental engineering law. It also includes studies on the environmental impact of proposed construction projects.

Environmental engineers study the effect of technological advances on the environment. To do so, they conduct studies on hazardous-waste management to evaluate the significance of such hazards, advice on treatment and containment, and develop regulations to prevent mishaps. Environmental engineers also design municipal water supply and industrial wastewater treatment systems as well as address local and worldwide environmental issues such as the effects of acid rain, global warming, ozone depletion, water pollution and air pollution from automobile exhausts and industrial sources.

At many universities, environmental engineering programs follow either the department of civil engineering or the department of chemical engineering at engineering faculties. Environmental "civil" engineers focus on hydrology, water resources management, bioremediation and water treatment plant design. Environmental "chemical" engineers, on the other hand, focus on environmental chemistry, advanced air and water treatment technologies and separation processes.

Additionally, engineers are more frequently obtaining specialized training in law and are utilizing their technical expertise in the practices of environmental engineering law.

Development

Ever since people first recognized that their health and well-being were related to the quality of their environment, they have applied thoughtful principles to attempt to improve the quality of their environment. The ancient Harappan civilization[①] utilized early sewers in some cities. Romans constructed aqueducts[②] to prevent drought and to create a clean, healthful water supply for the metropolis of Rome. In the 15th century, Bavaria[③] created laws restricting the development and degradation of alpine country that constituted the region's water supply.

The field emerged as a separate environmental discipline during the middle third of the 20th century in response to widespread public concern about water and pollution and increasingly extensive environmental quality degradation. However, its roots extend back to early efforts in public health engineering. Modern environmental engineering began in London in the mid-19th century when Joseph Bazalgette designed the first major sewerage system that reduced incidence of waterborne diseases such as cholera[④]. The introduction of drinking water treatment and sewage treatment in industrialized countries reduced waterborne diseases from leading causes of death to rarities.

In many cases, as societies grew, actions that were intended to achieve benefits for those societies had longer-term impacts which reduced other environmental qualities. One example is the widespread application of the pesticide DDT to control agricultural pests in the years following World War II. While the agricultural benefits were outstanding and crop yields increased dramatically, thus reducing world hunger substantially, and malaria was controlled better than it ever had been, numerous species were brought to the verge of extinction due to the impact of the DDT on their reproductive cycles. The story of DDT as vividly told in Rachel Carson's Silent Spring (1962) is considered to be the birth of the modern environmental movement and the development of the modern field of "environmental engineering".

Conservation movements and laws restricting public actions that would harm the environment have been developed by various societies for millennia. Notable examples are the laws decreeing the construction of sewers in London and Paris in the 19th century and the creation of the U.S. national park system in the early 20th century.

Words and Expressions

integration	[ˌɪntɪˈgreɪʃn]	n.	综合、结合、整合
engineering	[ˌendʒɪˈnɪərɪŋ]	n.	工程、工程学
habitation	[ˌhæbɪˈteɪʃn]	n.	居住、住所、生境
organism	[ˈɔːgənɪzəm]	n.	有机体、生物、（尤指）微生物
applied science			应用科学
plausible	[ˈplɔːzəbl]	adj.	似乎有理的、有道理的、可信的
adequate	[ˈædɪkwət]	adj.	充足的、适当的、足够的
sanitation	[ˌsænɪˈteɪʃn]	n.	卫生设备；卫生设施体系
disposal	[dɪˈspəʊzl]	n.	处置、清除、处理
hygiene	[ˈhaɪdʒiːn]	n.	卫生、卫生学
sustainability	[səsˌteɪnəˈbɪlɪti]	n.	持续性、永续性
propose	[prəˈpəʊz]	v.	提议、建议、打算
significance	[sɪgˈnɪfɪkəns]	n.	（尤指对将来有影响的）重要性、意义
regulation	[ˌregjuˈleɪʃn]	n.	章程、规章制度、规则、法规
ozone depletion			臭氧耗竭
automobile exhaust			汽车尾气
hydrology	[haɪˈdrɒlədʒɪ]	n.	水文学、水文地理学
process	[ˈprəʊses]	n.	（为达到某一目标的）过程、进程，（事物发展，尤指自然变化的）过程
specialized training			专业化训练
utilize	[ˈjuːtəlaɪz]	v.	使用、利用、运用、应用
sewer	[ˈsuːə(r)]	n.	污水管、下水道、阴沟
drought	[draʊt]	n.	干旱
metropolis	[məˈtrɒpəlɪs]	n.	大都会、大城市、首都、首府
incidence	[ˈɪnsɪdəns]	n.	发生范围、影响程度、发生率
cholera	[ˈkɒlərə]	n.	霍乱
sewage	[ˈsuːɪdʒ]	n.	（下水道的）污水
rarity	[ˈreərəti]	n.	珍品、稀有物、稀有、罕见
pesticide	[ˈpestɪsaɪd]	n.	农药、杀虫剂
malaria	[məˈleərɪə]	n.	疟疾
species	[ˈspiːʃiːz]	n.	种，物种（分类上小于属）
verge	[vɜːdʒ]	n.	边缘、路边、（道路的）植草边沿；v. 濒临、接近
extinction	[ɪkˈstɪŋkʃn]	n.	（植物、动物、生活方式等的）灭绝、绝种、消亡

conservation	[ˌkɒnsə'veɪʃn]	n.（对自然环境的）保护、防止流失（或浪费、损害、毁坏）、保持、保护
millennia	[mɪ'lenɪə]	n. 千年期、一千年
decree	[dɪ'kriː]	n. 法令、政令、（法院的）裁定、判决
		v. 裁定、判决、颁布

Notes

① Harappan civilization 哈拉帕文明，因其主要城市遗址哈拉帕得名。这种文化以印度河流域为中心，也称之为印度河流域文明。

② the Romans constructed aqueducts 罗马时期建成的引水渠，古罗马人在城市供水工程方面的成就突出。从公元前312至公元226年的500余年中，罗马城先后修建了11条大型输水道。

③ Bavaria 巴伐利亚，德国第二大州，首府慕尼黑。

④ waterborne disease 水传播疾病，如伤寒、疟疾、霍乱。

Exercises 1

1. According to the reading material, chose the best answer(s) from the options.

(1) Environmental engineering is the integration of ＿＿＿＿ to improve the natural environment.

  A. conservations and protections   B. sciences and engineering principles

  C. technologies and engineering principles   D. all the environmental sciences

(2) In the passage, waterborne diseases include＿＿＿.

  A. cholera   B. tuberculosis   C. malaria   D. stomachache

(3) Which of the following is NOT regarded as environmental engineering according to the passage?

  A. Study the effect of technological advances on the environment.

  B. Design municipal water supply.

  C. Evaluate the significance of hazardous wastes.

  D. Design the product.

(4) The pesticide most known to the general public is ＿＿＿.

  A. DDT   B. TNT   C. OPP   D. WP

(5) In the passage, the agricultural benefits does NOT mean ＿＿＿.

  A. crop yields increased

  B. reducing world hunger substantially

  C. malaria was controlled better than it ever had been

D. numerous species were brought to the verge of extinction

2. Give some examples of conservation movements.

3. List scopes of environmental engineering based on your understanding of the passage.

Part 2　Translation

科技英语的特点（上）

随着现代科学技术的发展，科技英语已发展成为一种重要的英语文体，与新闻报刊文体、论述文体、公文文体、描述及叙述文体等应用文体一道，形成了英语中六大主要文体。总体来说，科技英语文体的特点可以通过其词汇、句法、修辞、语篇等层面表现出来。

一、科技英语的词汇特点

1. 多用术语，专业性强

科技英语的内容都与具体的科技领域息息相关，每一门学科或专业都有一定数量的术语和词汇的习惯用法。专业术语能正确表达科学概念，具有丰富的内涵和外延。专业性强是科技英语词汇的基本特点。为了揭示自然科学和客观事物的现象和发展规律，科技英语必须使用表义确切的专业术语。例如，eutrophication（富营养化）、denitrification（反硝化作用）等词汇，一看便知是专业性极强的词汇。

2. 一词多义，内涵外延

同样一个英语词汇，它的普通意义和专业术语意义相去甚远，有时是风马牛不相及。例如，fault 的基本词义为"错误"，但在地质学中意为"断层"，在电子学中意为"电路故障"，在排球、篮球等体育运动中意为"传球失误"。又如，eye 一般意为"眼睛"，但在科技文献中，根据不同的专业和语境其涵义也不尽相同，如 inspection eye（检查孔）、hoisting eye（吊环）、observers eye（观测透镜）、deal eye（衬圈）、television eye（工业电视摄像机）等。对于这类一词多义的现象，只有通过该词使用的语境加以判别。

3. 构词灵活，表达简练

科技英语词汇构词方法多样，有合成法、借用法、缩略法、派生词、混合法、旧词新义法和杜撰新词法等，其中最常见的是合成法、派生法和混合法。

合成法就是将两个或两个以上的词合在一起构成一个新词。例如 hairline（游丝、细测量线）、splashdown（溅落）。此外，还有些合成词需要有连字符连接才能构成，例如，pulse-scaler（脉冲定标器）、baby-blues（产后抑郁症）等。

派生法是用词缀和词根结合构成新词。英语中有大量的词缀,由于其强大的附着力与大量的词根相组合便会产生无数的新概念。一般也可分为前缀和后缀两大类,例如 semisphere(半球)(semi 为前缀,意思为"一半")、metropolis(大都市)(polis 作为后缀,表示"国家或城市")、recycle(再循环)(前缀 re 的意思为"再一次,重新")。

混合法是将词语缩略以后合成新词,换句话说,也就是将原有两词各取其中一部分(有时还可以是某个词的全部)。不过,这类新词还保留原有词汇的某些特征,读者很容易区分开来。例如 kidult(拒绝长大的成年人)(由 kid 和 adult 缩合而成)、telethon(煲电话粥)(由 telephone 和 marathon 缩写而成)、technoburb(科技园区)(由 technology 和 exurb 缩合而成)。

4. 新词不断,语义更新

随着科技的日新月异,新事物、新概念和新现象大量涌现,语言中记录这些变化的新词也越来越多,如 stereo-chemistry(立体化学)、bioinformatics(生物信息学)、superconducting material(超导材料)等。

二、科技英语的句法特点

1. 多用长难句

科技英语中常常需要表达多重密切相关的概念,这些严谨的逻辑和严密的阐述常常是通过运用有复杂句法结构的长句子实现的。一般来说,长难句的结构特点是后置定语、非谓语动词、同位语、宾语从句、定语从句、状语从句等成分多,有时一个句子能包含所有这些成分。

长句表达虽有条理性、周密性和严谨性等优点,但其缺点就是因为句中包含太多修饰成分、限制成分和短语,给翻译带来很大困难。因此,翻译长难句时需要注意:①弄清句子的逻辑关系;②根据上下文和全句内容领会句子要义;③辨别该长句的主从结构,分切句子的内容;④分清上下层次及前后联系,然后根据汉语的特点、习惯和表达方式翻译。例如:

The absence of pesticides and the emphasis on natural fertilizer are designed not only to keep the experiment as untainted as possible, but also to protect the health of the human consumers; because all the air and water in Biosphere Ⅱ is continually recycled and regenerated, it is important that no poisons be introduced into the system anywhere.

译文:不使用农药和强调天然肥料的目的不仅是使实验尽可能不被污染,也是保护食用者的健康。因为"第二生物圈"的空气和水全部是不断地再循环和再生的,所以不在此生态系统中任何地方引入毒物是很重要的。

原文由两大部分组成,中间由分号隔开。分号前一部分为一个分句,其中包含有 not only...but also 连接的两个不定式短语 to keep 与 to protect;分号后面的部分为一个主从复合句,由 because 引导的原因状语从句和 it 做形式主语的主句所组成,真正的主语是 that 引导的主语从句。弄清楚这个长句中层层堆叠的语法结构对于理解整个句子至关重要。

2. 多用名词化结构

名词化是指词性作用的名词性转化，如起名词作用的非谓语动词和与动词同根或同形的名词，也包括一些形容词来源的名词。这些词可起名词的作用，也可表达谓语动词或形容词所表达的内容，常伴有修饰成分或附加成分，构成短语。这种短语称为名词化结构。

名词化结构的组合方式多，意义容量大，适宜于表达精细复杂的思想，使文章具有庄重感和严肃感。大量使用名词化结构符合科技文体的要求，使语体更加正式、更具书面语风格。例如：

(1) Those who moved to colder climates developed light skin to take advantage of the sunlight for the *synthesis* of vitamin D.

译文：迁移到较冷地带的人生成较白的皮肤，以利用日光来合成维生素 D。

例（1）中 synthesis 由动词 synthesize 名词化而来，从而把物质过程名词化，表示"合成"这一过程。

(2) The screw is also used in calipers for *measuring* very accurately the diameters of small objects.

译文：此螺钉也可用在用以精密测量小物件直径的卡钳上。

例（2）中动名词 measuring 除了带有宾语，还带有状语成分 very accurately，动名词短语用作介词 for 的宾语。

3. 多用非谓语动词形式

英语中动词可分为限定（finite）和非限定（non-finite）两类。英语中的非谓语结构就是不用作谓语的动词或动词短语，因为这种结构不受主语的限定，没有人称和数的变化，因此也称为非限定动词或非谓语动词。非谓语动词分三种：不定式动词、分词（现在分词和过去分词）、动名词。

由于科技英语在客观上要求语义明确、表达简练、结构紧凑，所以非谓语动词（即非限定动词）的使用频率要比一般英语文体高得多。一方面，为了使描述的对象更加明确，常需要用非谓语动词做定语加以限定；另一方面，为了使语句简练，常需要用非谓语动词短语来代替各种从句，特别是定语从句和状语从句。例如：

Environmental engineering is the integration of sciences and engineering principles *to improve* the natural environment, *to provide* healthy water, air and land for human habitation and other organisms, and *to clean up* pollution sites.

译文：环境工程是科学和工程原理的结合，可以改善自然环境，为人类和其他生物的栖息提供健康的水、空气和土地，并清理污染场所。

上例就是用不定式动词替代定语从句，从而使句子结构简洁，避免复杂的从句句式。

4. 多用定语（从句）结构

定语从句是英语从句中最为复杂的一种。由于科技英语的文体特征是概念清晰、逻辑性强，因而常用能够表达完整意思的定语从句修饰名词或代词，以便明晰、确切地表达该词的概念，因此，定语从句在科技英语中大量使用。定语从句不仅结构复杂，而且含义繁多，具有补充、原因、转折、结果、目的、条件、让步等意义。按照定语在句中的位置，可分为前置定语和后置定语，后置定语还可能环环相扣，一个接一个，使句子结构纷繁复杂，给翻译带来了巨大困难。例如：

The gradual development of these building blocks, following the invention of the roller or wheel and the arm or lever in ancient times, brought about the Industrial Revolution, starting with the assembly of James Watt's *engine* for harnessing the force of steam and proceeding into the advanced *mechanization* of present automatic control.

译文1：随着远古轧辊、轮子及臂状物或杠杆的发明，这些构建块逐渐发展起来，并带来了工业革命，并以詹姆斯·瓦特的以蒸汽为动力的发动机的组装为开始，现在已进入高级机械自动化的阶段

译文2：随着远古轧辊、轮子及臂状物或杠杆的发明，这些构建块逐渐得到发展从而产生了工业革命——工业革命开始于詹姆斯·瓦特发明的以蒸汽机为动力的发动机，现在已进入自动控制这种高级机械化阶段。

上例中心词 engine 既有名词所有格形式的前置定语，也有介词短语做后置定语。同样，中心词 mechanization 既有形容词做前置定语，又有介词短语做后置定语，而且介词短语中的中心词还有两个形容词做前置定语。如果按照源语言语序进行翻译，会导致译文逻辑关系不清、表达不畅。比较译文1和译文2，译文2则为佳译。

5. 多采用惯用的句型结构

科技英语中，经常采用一些惯用句型，如以 it 为形式主语的主语从句和不定式短语做主语（如 It seems that…/ It is reported that…/ It has been proved that…）、带有表语或表语从句的陈述句型（如 The fact is that…/ The truth is that…/ Of recent concern is…）、含有宾语或宾语从句的陈述句（如 This means that…/Experiment shows that …/Practice has proved that…）、带有条件状语从句的句型（如 If…/When…）。例如：

The truth is that the current increases with every decease of resistance.

译文：真实情况是：电阻一减小，电流就增加。

上例中的表语从句用 that 做从属连接词，引导表语从句，但 that 没有实际意义，只起连接作用，故不必译出。

6. 多采用省略结构

科技英语中，信息明晰和表达简练是基本要求，所以科技英语句子写作中要尽量杜绝不必要的重复。同时，为了增加句子信息容量，使句子结构紧凑并富于变化，有必要省略句中不必要的"冗余"信息。省略的部分可以是一个词、一个短语等。例如：

In electrolysis the water breaks up into hydrogen and oxygen, the hydrogen appearing about the cathode and oxygen about the anode.

译文：水电解时分解为氢和氧，氢出现在负极周围，而氧出现在正极周围。

上例中，短语 the hydrogen appearing about the cathode 和短语 oxygen about the anode 是并列成分，短语 oxygen about the anode 中省略了 appearing。

在长期的使用过程中，科技英语形成了一些形式固定的惯用省略句型，大都由状语从句缩略而来。例如：

If necessary（如有必要，必要时）；
If possible（如有可能）；
When needed（需要时，如有需要）；
Where feasible（如果可行的话）；
As noted later（如后所述，从下文可以看出）；
As previously mentioned（前已提及，如前文所述）；

这类的惯用法很多，同学们可以多加留意。

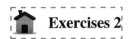

Exercises 2

1. Translate the following sentences into Chinese.

(1) Environmental engineering is the integration of sciences and engineering principles to improve the natural environment, to provide healthy water, air and land for human habitation and other organisms, and to clean up pollution sites.

(2) Furthermore, it is concerned with finding plausible solutions in the field of public health, such as waterborne diseases, implementing laws which promote adequate sanitation in urban, rural and recreational areas.

(3) In many cases, as societies grew, actions that were intended to achieve benefits for those societies had longer-term impacts which reduced other environmental qualities.

2. Translate the following sentences into English.

（1）DDT 在为农业做出巨大贡献时，也为整个生态系统带来了前所未有的灾难。

（2）如有可能，环境工程师会不遗余力地恢复功能受损的环境，减少对环境产生威胁的人类活动，因为没有人比他们更了解环境的重要性。

Part 3　Writing

科技英语是英语语言诸多变体中的一种，它与常见的文学语言表述有明显的不同。它用词客观、直叙、简练、准确，所叙述的过程具有很强的可操作性，而且它常用行为语言对事物及过程进行描述。这些特点都是科技英语写作中需要特别注意的。

一、科技英语写作的主要类型

1. 科技信息类

科技信息类是指将已出版或发行的英语科技论文、专著进行摘要，然后归类储存，以备在科研或实际工作中使用。例如，美国的 CI（美国化学索引）、EI（美国工程技术索引）等就是属于科技信息类写作，这项工作往往由专业人士承担。

2. 科技报告类

科技报告类通常是对一件科技活动的高度概括，把科技活动所涉及的时间、地点、人员、目的、主要问题以及应对方法或建议等以文字的形式提供给接收方。主要包括实验报告、事故报告、进展报告、可行性报告和参会报告等。

3. 科技论文类

科技论文是将作者在某一学术领域内就某一课题所研究出的成果和结果以及取得这些成果的理论依据、方法、步骤、所使用的工具等以文字的形式提供给读者。科技论文主要有以下四种：毕业论文（graduation thesis）、学位论文（degree thesis）、投稿论文（contributed thesis）和预约论文（assigned thesis）。

4. 设计说明书类

设计说明书是说明某一物品的设计原理、技术参数、使用注意事项、操作该物品的行为步骤，例如：如何安装、使用、检查、排除故障，如何进行维修、检测，以及在实验室里如何工作等。

5. 科技应用文类

科技应用文是指科技工作者之间来往的业务信函、专利申请、求职信函等。

二、科技英语写作的要求

1. 准确（accuracy）

准确的概念体现在内容与格式两个方面。内容准确就是要表达准确，要正确地运用英语的语法与句型。在科技英语写作中，尤其是在进行过程描述和定义等的写作中，正确应用英语语法句型和词汇显得尤为重要。

【例1】

A. People commonly say that clinical thermometers will be used to find out the body temperature.（句型使用错误，并且使用了错误的时态）

B. It is commonly believed that clinical thermometers are used to determine the body temperature.（句型与时态均正确）

从上述两个例句中可以清楚地看出，第一句犯了中国人学外语常犯的错误，并且在表示科学真理时，使用了错误的时态。在第二句中，使用了符合英语表达习惯的句型，并且使用了符合科技英语在表示科学真理时要使用一般现在时或一般过去时的要求，故为正确的句子。

另外，不要使用意义模棱两可的词汇和表达方式。在科技英语写作中，对这一问题尤需注意。特别是在下定义、描述过程、写操作说明书、做结论等时，必须避免使用诸如 might be、possible、probable 等这些表示意义不确定的词汇和表达方式，而应当使用意义清楚、表达准确的词汇。

使用正确的格式也是衡量科技英语准确性的一个重要标准。格式的准确性包括两个方面的内容：

（1）在表示定义、过程描述、操作说明书、做结论等的表述中，不能使用带有主观色彩的词汇，如 I think、 in my mind、 I guess、 I imagine 这类词汇。

（2）科技英语的文章要严格按照科技英语不同文体的格式来写作。比如写摘要时，不能加入个人观点，也不要用问句和感叹句，最好使用第三人称。

2. 简洁（brevity/ conciseness）

任何文章都应简洁，科技文章对此则尤为强调。科技文章写作的内容通常包括研究的目的和范围、研究的方法、研究的步骤、研究的结果和作者对所进行的研究的结论。因此，科技英语写作在文体上所要求的简洁主要基于下列三个因素：

（1）节省时间。如果用简洁的文字可以把想表达的思想写出来，就可以节省写作的时间，同时也可以节省阅读的时间。

（2）简洁的文字有利于读者对文字的理解。

（3）科技英语文体的文章是有篇幅限制的，以写个人简历为例，长度最好不超过一页。掌上电脑或者便携式电脑，或者手机，它们的屏幕都不大，通常不能显示整页。所以，使用简洁语言的重要性会更加突显。

3. 清楚（clarity）

一篇优秀的科技文章的基本要求就是表达清楚。表达清楚首先在于作者思维的逻辑性，其次在于表达的连贯性。为了在科技英语写作中达到内容表达清楚的要求，以下这些原则必须遵循：

（1）提供明确的信息。在科技英语写作中实现内容表达清楚的一个方法就是使用明确的或者具体量化的信息。如果使用抽象的、概念模糊的形容词或副词，如 some、 recently 这类词汇，读者就可能会对文字产生不同的理解。

（2）回复必须简明扼要。对收到的报告中的问题进行回复时，针对性要强，特别是基本信息要明确，如对人、事、时间、地点、原因和方法的回答要具体。

（3）使用容易理解的词汇。表达清楚的另一个关键是使用读者容易理解的词汇，要避免使用那些意义模糊的词汇，特别是当使用缩略语和专业术语时一定要仔细。

【例2】

A. ACTIVITY REPORT DRAFT

Our <u>latest</u> attempt at molding performance protectors has led to <u>some</u> positive results. We spent <u>several</u> hours in Dept.15 trying different machine settings and techniques. <u>Several</u> good parts were molded using two different sheet thicknesses. Here is a summary of the findings. First we tried the <u>thick</u> sheet material. At 240°F, this thickness worked well.

Next, we tried the thinner sheet material. The <u>thinner</u> material is less forgiving, but after a <u>few</u> adjustments we were making good parts. Still, the <u>thin</u> material caused the most handling problems.

在 A 中作者使用了一些概念表达比较模糊的词汇，这是科技英语写作中应该避免的。为了完善这些表达不清楚的地方（有下划线的词汇），对上面的这份报告进行修改，特别是对意义模糊的形容词进行了量化。B 就是经过修改的报告。

B. ACTIVITY REPORT DRAFT

During the week of 19/8/05, we spent approximately 12 hours in Dept.15 trying different machine settings, techniques, and thicknesses to mold performance and mold protectors. Here is a report of our findings.

<u>030# Thick Sheet</u>

At 240°F, this thickness worked well.

<u>015# Thick Sheet</u>

This material is less forgiving, but after decreasing the heat to 200°F, we could produce good parts. Still, material at 015# causes handling problems.

在 B 中，作者使用了具体的数字，使得表达清楚，达到了交流的目的。

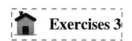

Exercises 3

Summarize the reading material with no less than 150 words, based on the requirements of scientific English writing.

Expanding Reading

Job Prospects of Environmental Engineer

Environmental engineers use knowledge of engineering, soil science, chemistry, and biology to solve problems in theenvironment. They tackle a variety of issues, and their concerns include pollution control, recycling, and public health issues.

For example, an environmental engineer might work on devising solutions for effective wastewater management. This could include designing systems to treat industrial wastewater, manage municipal water supply, prevent waterborne diseases, and improve the sanitation in cities, recreational areas, and rural locations.

Duties

Some of the duties and responsibilities an environmental engineer might engage in can include the following:
- Make recommendations to maintain and improve environmental performance.
- Review environmental regulations, and determine whether they're being applied properly.
- Review stormwater management practices for municipal, industrial, and construction stormwater programs.
- Create and maintain air quality management systems that comply with air permits and air regulations.
- Report environmental incidents to plant management, including mishaps such as internal spills, external releases, potential permit non-compliances, and upcoming regulatory inspections.
- Lead or support the preparation and negotiation of various environmental permit applications.
- Interface with different regulatory agencies, prepare needed documentation, schedule required testing, and provide any necessary, additional follow-up documentation.

Payment

The median annual wage for environmental engineers was $87,620 in May 2018. The median wage is the wage at which half the workers in an occupation earned more than that amount and half earned less. The lowest 10 percent earned less than $53,180, and the highest 10 percent earned more than $137,090 (the median wage of all occupations in the U.S. is $38,640). In May 2018, the median annual wages for environmental engineers in the top industries in which they worked were as follows:

Federal government, excluding postal service	$105,410
Local government, excluding education and hospitals	$87,910
Engineering services	$86,670
Management, scientific, and technical consulting services	$81,110
State government, excluding education and hospitals	$80,370

Most environmental engineers work full time. Those who manage projects often work more than 40 hours per week to monitor the project's progress, ensure that deadlines are met, and recommend corrective action when needed.

(*From U.S. Bureau of Labor Statistics, Occupational Employment Statistics*)

(*https://www.thebalancecareers.com/environmental-engineer-526013*)

UNIT 1 Introduction to Environmental Engineering

Lesson 2 Global Environmental Concerns

Part 1 Reading

As early as 1896, the Swedish scientist Svante Arrhenius had predicted that human activities would interfere with the way the sun interacts with the earth, resulting in global warming and climate change. His prediction has become true and climate change is now disrupting global environmental stability. The last few decades have seen many treaties, conventions, and protocols① for the cause of global environmental protection.

Few examples of environmental issues of global significance are:
- Ozone layer depletion.
- Global warming.
- Loss of biodiversity.

One of the most important characteristics of this environmental degradation is that it affects all mankind on a global scale without regard to any particular country, region, or race. The whole world is a stakeholder and this raises issues on who should do what to combat environmental degradation.

1 Ozone Layer Depletion

Earth's atmosphere is divided into three regions, namely troposphere, stratosphere and mesosphere. The stratosphere extends from 10 to 50 km from the Earth's surface. This region is concentrated with slightly pungent smelling, light bluish ozone gas. The ozone gas is made up of molecules each containing three atoms of oxygen; its chemical formula is O_3. The ozone layer, in the stratosphere acts as an efficient filter for harmful solar Ultraviolet B (UV-B) rays. Ozone is produced and destroyed naturally in the atmosphere and until recently, this resulted in a well-balanced equilibrium. Ozone is formed when oxygen molecules absorb ultraviolet radiation with wavelengths less than 240 nanometres and is destroyed when it absorbs ultraviolet radiation with wavelengths greater than 290 nanometres. In recent years, scientists have measured a seasonal thinning of the ozone layer primarily at the South Pole. This phenomenon is being called the ozone hole.

1.1 Effects of Ozone Layer Depletion

Effects on Human and Animal Health: Increased penetration of solar UV-B radiation is likely to have high impact on human health with potential risks of eye diseases, skin cancer and infectious diseases.

Effects on Terrestrial Plants: In forests and grasslands, increased radiation is likely to change species composition thus altering the bio-diversity in different ecosystems. It could also affect the plant community indirectly resulting in changes in plant form, secondary metabolism, etc.

Effects on Aquatic Ecosystems: High levels of radiation exposure in tropics and subtropics may affect the distribution of phytoplanktons, which form the foundation of aquatic food webs. It can also cause damage to early development stages of fish, shrimp, crab, amphibians and other animals, the most severe effects being decreased reproductive capacity and impaired larval development.

Effects on Bio-geo-chemical Cycles: Increased solar UV radiation could affect terrestrial and aquatic bio-geo-chemical cycles thus altering both sources and sinks of greenhouse and important trace gases, e.g. carbon dioxide (CO_2), carbon monoxide (CO), carbonyl sulfide (COS), etc. These changes would contribute to biosphere-atmosphere feedbacks responsible for the atmosphere build-up of these greenhouse gases.

Effects on Air Quality: Reduction of stratospheric ozone and increased penetration of UV-B radiation result in higher photo dissociation rates of key trace gases that control the chemical reactivity of the troposphere. This can increase both production and destruction of ozone and related oxidants such as hydrogen peroxide, which are known to have adverse effects on human health, terrestrial plants and outdoor materials.

The ozone layer, therefore, is highly beneficial to plant and animal life on earth filtering the dangerous part of sun's radiation and allowing only the beneficial part to reach earth. Any disturbance or depletion of this layer would result in an increase of harmful radiation reaching the earth's surface leading to dangerous consequences.

1.2 Ozone Depletion Counter Measures

- International cooperation, agreement (Montreal Protocol[2]) to phase out ozone depleting chemicals since 1974.
- Tax imposed for ozone depleting substances.
- Recycle of CFCs[3] and Halons[4].

2 Global Warming

Before the Industrial Revolution, human activities released very few gases into the atmosphere and all climate changes happened naturally. After the Industrial Revolution, through fossil fuel combustion, changing agricultural practices and deforestation, the natural composition of gases in the atmosphere is getting affected and climate and environment began to alter significantly.

Over the last 100 years, it was found out that the earth is getting warmer and warmer, unlike previous 8 000 years when temperatures have been relatively constant. The present temperature is 0.3-0.6℃ warmer than it was 100 years ago. The key greenhouse gases (GHG) causing global warming is carbon dioxide. CFCs, even though they exist in very small quantities, are significant contributors to global warming. Carbon dioxide, one of the most prevalent greenhouse gases in the atmosphere, has two major anthropogenic (human-caused) sources: the combustion of fossil fuels and changes in land use. Net releases of carbon dioxide from these two sources are believed to be contributing to the rapid rise in atmospheric concentrations since Industrial Revolution. Because estimates indicate that approximately 80 percent of all anthropogenic carbon dioxide emissions currently come from fossil fuel combustion, world energy use has emerged at the center of the climate change debate.

2.1 Sources of Greenhouse Gases

Some greenhouse gases occur naturally in the atmosphere, while others result from human activities. Naturally occurring greenhouse gases include water vapor, carbon dioxide, methane, nitrous oxide, and ozone. Certain human activities, however, add to the levels of most of these naturally occurring gases. Carbon dioxide is released to the atmosphere when solid waste, fossil fuels (oil, natural gas, and coal), and wood and wood products are burned. Methane is emitted during the production and transport of coal, natural gas, and oil. Methane emissions also result from the decomposition of organic wastes in municipal solid waste landfills, and the raising of livestock. Nitrous oxide is emitted during agricultural and industrial activities, as well as during combustion of solid waste and fossil fuels.

Very powerful greenhouse gases that are not naturally occurring include hydrofluorocarbons (HFCs), perfluorocarbons (PFCs), and sulfur hexafluoride (SF_6), which are generated in a variety of industrial processes. Often, estimates of greenhouse gas emissions are presented in units of millions of metric tons of carbon equivalents (MMTCE), which weights each gas by its Global Warming Potential or GWP value.

2.2 Global Warming (Climate Change) Implications

2.2.1 Rise in global temperature

Observations show that global temperatures have risen by about 0.6℃ over the 20th century. There is strong evidence now that most of the observed warming over the last 50 years is caused by human activities. Climate models predict that the global temperature will rise by about 6℃ by the year 2100.

2.2.2 Rise in sea level

In general, the faster the climate changes, the greater will be the risk of damage. The mean sea level is expected to rise 9-88 cm by the year 2100, causing flooding of low lying areas and other damages.

2.2.3 Food shortages and hunger

Water resources will be affected as precipitation and evaporation patterns change around the world. This will affect agricultural output. Food security is likely to be threatened and some regions are likely to experience food shortages and hunger.

3　Loss of Biodiversity

Biodiversity refers to the variety of life on earth, and its biological diversity. The number of species of plants, animals, microorganisms, the enormous diversity of genes in these species, the different ecosystems on the planet, such as deserts, rainforests and coral reefs are all a part of a biologically diverse earth. Biodiversity actually boosts ecosystem productivity where each species, no matter how small, all have an important role to play and that it is in this combination that enables the ecosystem to possess the ability to prevent and recover from a variety of disasters.

It is now believed that human activity is changing biodiversity and causing massive extinctions. The World Resource Institute reports that there is a link between biodiversity and climate change. Rapid global warming can affect ecosystems chances to adapt naturally. Over the past 150 years, deforestation has contributed an estimated 30 percent of the atmospheric build-up of CO_2. It is also a significant driving force behind the loss of genes, species, and critical ecosystem services.

Words and Expressions

interfere	[ˌɪntəˈfɪə(r)]	v. 干涉、干预、介入
interact	[ˌɪntərˈækt]	v. 互相影响、互相作用、交流、沟通
disrupt	[dɪsˈrʌpt]	v. 破坏、使瓦解、扰乱、使中断
treaty	[ˈtriːtɪ]	n.（国家之间的）条约、协议
convention	[kənˈvenʃn]	n.（国家或首脑间的）公约、习俗、常规、惯例
stakeholder	[ˈsteɪkhəʊldə(r)]	n.（某组织、工程、体系等的）参与人、参与方、有权益关系者
ozone layer depletion		臭氧层耗竭

biodiversity	[ˌbaɪəʊdaɪˈvɜːsətɪ]	n. 生物多样性
combat	[ˈkɒmbæt]	n. 搏斗、打仗、战斗； v. 防止、减轻、战斗、与……搏斗
atmosphere	[ˈætməsfɪə(r)]	n. 大气，大气层、大气圈
troposphere	[ˈtrɒpəsfɪə(r)]	n.（大气层的）对流层
stratosphere	[ˈstrætəsfɪə(r)]	n.（大气层的）平流层
mesosphere	[ˈmezəsfɪə(r)]	n.（大气层的）中间层
pungent	[ˈpʌndʒənt]	adj.（气味）强烈的、刺激性的
bluish	[ˈbluːɪʃ]	adj. 带蓝色的、有点蓝的
molecule	[ˈmɒlɪkjuːl]	n. 分子
atom	[ˈætəm]	n. 原子
formula	[ˈfɔːmjələ]	n. 公式、方程式、计算式、分子式
equilibrium	[ˌiːkwɪˈlɪbriəm]	n. 平衡、均衡、均势
filter	[ˈfɪltə(r)]	n. 过滤器、筛选； v. 过滤
ultraviolet	[ˌʌltrəˈvaɪələt]	n. 紫外线辐射； adj. 紫外线的、利用紫外线的
radiation	[ˌreɪdɪˈeɪʃn]	n. 辐射、放射线、辐射的热（或能量等）
wavelength	[ˈweɪvleŋθ]	n. 波长、（广播电台等占用的）频道
nanometer	[ˈnænəʊmiːtə(r)]	n. 纳米
absorb	[əbˈzɔːb]	v. 吸收（液体、气体等）
penetration	[ˌpenəˈtreɪʃn]	n. 渗透、穿透
impact	[ˈɪmpækt, ɪmˈpækt]	n. 巨大影响、强大作用、撞击、冲撞、冲击力； v.（对某事物）有影响、有作用、冲击、撞击
potential	[pəˈtenʃl]	n. 潜能、可能性、潜在性、电位、电势、电压； adj. 潜在的、可能的
composition	[ˌkɒmpəˈzɪʃn]	n. 成分、构成、组合方式
ecosystem	[ˈiːkəʊsɪstəm]	n. 生态系统
community	[kəˈmjuːnətɪ]	n. 群落、社区
metabolism	[məˈtæbəlɪzəm]	n. 新陈代谢
secondary	[ˈsekəndrɪ]	adj. 第二的、次要的、从属的、辅助的、间接引发的、继发性的
exposure	[ɪkˈspəʊʒə(r)]	n. 面临、遭受（危险或不快）、揭

		露、被报道
tropic	[ˈtrɒpɪk]	n. 热带
distribution	[ˌdɪstrɪˈbjuːʃn]	n. 分布、分配、分发、分送
phytoplankton	[ˌfaɪtoʊˈplæŋtən]	n. 浮游植物、藻类
reproductive capacity		生殖能力
impair	[ɪmˈpeə(r)]	v. 损害、削弱
larval	[ˈlɑːvl]	adj. 幼虫的、幼体的
terrestrial	[təˈrestrɪəl]	n. 陆地生物;
		adj. 陆地的
carbonyl sulfide		硫化羰、羰基硫化物
feedback	[ˈfiːdbæk]	n. 反馈的意见（或信息）
dissociation	[dɪˌsəʊʃɪˈeɪʃn]	n. 分解、分离、分裂
destruction	[dɪˈstrʌkʃn]	n. 破坏、毁灭、摧毁
oxidant	[ˈɒksɪdənt]	n. 氧化剂
hydrogen peroxide		过氧化氢
substitute	[ˈsʌbstɪtjuːt]	n. 替代品
halon		哈龙
deforestation	[ˌdiːˌfɒrɪˈsteɪʃn]	n. 采伐森林、森林开伐
constant	[ˈkɒnstənt]	n. 常数、常量;
		adj. 连续发生的、不断的、不变的、固定的、恒定的
prevalent	[ˈprevələnt]	n. 流行的、普遍的、盛行的
combustion	[kəmˈbʌstʃən]	n. 燃烧、燃烧过程
methane	[ˈmiːθeɪn]	n. 甲烷、沼气固体废物
solid waste		固体废弃物
decomposition	[ˌdiːˌkɒmpəˈzɪʃn]	n. 分解、腐烂、变质
organic wastes		有机废物
landfill	[ˈlændfɪl]	n. 垃圾填埋场
hydrofluorocarbon	[ˌhaɪdrəʊˈflʊərəʊkɑːbən]	n. 氢氟烃
perfluorocarbon	[ˌpəˈflʊərəʊkɑːbən]	n. 全氟化碳
equivalent	[ɪˈkwɪvələnt]	n. 等价物、等量;
		adj. （价值、数量、意义、重要性等）等价的、相等的
sulfur hexafluoride		六氟化硫
precipitation	[prɪˌsɪpɪˈteɪʃn]	n. 降水、降水量（包括雨、雪、冰等）沉淀
evaporation	[ɪˌvæpəˈreɪʃn]	n. 蒸发、挥发

| gene | [dʒiːn] | *n.* 基因 |
| coral reefs | | 珊瑚礁 |

 Notes

① Treaty, Convention, Protocol 条约、公约、议定书，具有约束力的国际协议，treaty 是双边协议，convention 是多边协议，protocol 是指缔约国对条约公约的解释、补充、修改所议定缔结的国际法律文件。

② Montreal Protocol 全称是 The Montreal Protocol on Substances that Deplete the Ozone Layer,《关于耗损臭氧层物质的蒙特利尔议定书》。

③ CFC 氯氟烃，俗称氟利昂，破坏臭氧层的物质，全称是 Chlorofluoroncarbon

④ Halon 哈龙，它属于一类称为卤代烷的化学品，主要用于灭火药剂，中国受控消耗臭氧层物质清单列为第二类。

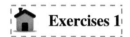 **Exercises 1**

1. According to the reading material, chose the best answer(s) from the options.

(1) Which of the following is NOT described as global environmental issues?_____

A. Ozone Layer Depletion   B. Loss of Biodiversity

C. Global Warming   D. Photochemical Smog

(2) The role of ozone layer most known to the public is its_____.

A. reflection of electromagnetic waves   B. absorbtion of sun's heat

C. absorbtion of harmful sun's radiation   D. fall kinds of weather

(3) The three forms of international law are _____.

A. treaty   B. convention   C. regulation   D. protocol

(4) CFC and Halons are the key contributors to _____.

A. Loss of Biodiversity   B. Ozone Layer Depletion

C. Global Warming   D. Photochemical Smog

(5) According to the passage, _____ and _____ also make contributions to methane emissions.

A. decomposition of organic waste   B. agricultural activities

C. combustion of fossil fuel   D. raising livestock

2. Describe the greenhouse effect in English.

3. Discuss the role of biodiversity in English.

Part 2　Translation

科技英语的特点（下）

一、科技英语的修辞特点

表达特定内容的文体，有与之相适应的修辞方式。科技英语中的修辞特点主要体现在时态、语态、语气的运用等方面。

1. 时　态

科技英语中常用的有一般现在时、一般过去时、现在完成时和一般将来时四种时态。每种时态的使用都有其基本要求和适应范围。使用最多的是一般现在时，它常用于对科学定义、定理、方程（或公式）或图表等进行解说，这些客观真理性的内容是没有时间性的。此外，一般现在时可用于表述经常发生的或没有时限的自然现象、过程和规则。

例如：

(1) Nylon is nearly twice as strong and much less affected by water than natural silk.

译文：尼龙的强度几乎是天然丝的两倍，且不像天然丝那样易受水的影响。

(2) One of the most important characteristics of this environmental degradation is that it affects all mankind on a global scale without regard to any particular country, region, or race.

译文：这种环境恶化的最主要的特征之一就是它在全球尺度上影响全人类，无论哪一个国家、地区或种族都不能幸免。

科技英语中，在表示过去的时间或叙述过去发生过的事实的情况时，如陈述已经发生的自然现象、已经做过的实验、已经从事过的研究或者已经进行的活动，常采用一般过去时态。例如：

(3) Early fires on the earth were certainly caused by nature, not by Man.

译文：地球上早期的火肯定是由大自然而不是人类引燃的。

(4) Over the last 100 years, it was found out that the earth is getting warmer and warmer, unlike previous 8 000 years when temperatures have been relatively constant.

译文：在过去的100年里，地球变得越来越温暖，这有异于在更早的8000年中温度一直相对稳定的情况。

需要注意的是，在叙述历史事件或介绍事物的历史发展过程时，也需要使用一般过去时，如例（4）。

在表示已经发现或获得的研究成果时，若与现在关系直接且仍有影响则要用现在完成时。过去发生的事情可以是事件，也可以是一种状态，但无需指明事情发生的具体时间。例如：

(5) His prediction has become true and climate change is now disrupting global

environmental stability.

译文：他的预言现已成真，气候变化正扰动着全球的环境稳定性。

当需要陈述对未来工作安排、技术活动、研究展望或者结果预测时，常使用一般将来时。例如：

(6) Water resources will be affected as precipitation and evaporation patterns change around the world.

译文：全球的降水和蒸发模式的变化会影响到水资源。

过去完成时常配合一般过去时、现在完成时等使用。例如：

(7) As early as 1896, the Swedish scientist Svante Arrhenius had predicted that human activities would interfere with the way the sun interacts with the earth, resulting in global warming and climate change.

译文：早在 1896 年，瑞典科学家 Svante Arrhenius 就预言，人类活动会影响太阳对地球的照射，从而导致全球变暖和气候变化。

在科技英语中，时态的使用也可以是丰富灵活的，可以将多种时态一起使用，而不是害怕出错就刻板地使用一种时态。例如，阅读资料中的第一段，就使用了过去完成时和一般完成时，用来表达科学界对环境问题长期以来的关注以及对一些已经被预见的问题仍然存在表达了遗憾，这样也会为我们准确、简洁、清楚地表达增色不少。

2. 语　态

科技英语中被动语态是最常用的。据统计，在物理、化学、工程类英文材料里，全部限定动词中至少有 1/3 是被动语态。这是因为科技文章反映的是客观事实及据此做出的科学推论，因此语言运用要体现客观性和普遍性，避免主观臆断。此外，科技文章描述的是科学研究的对象、手段、过程结果等各个方面，揭示客观世界的规律，使用被动语态可以突出客观世界这一科学研究的主体。

被动语态一般由"助动词'be'+动词的过去分词"或"get/ become+动词的过去分词"构成。一般来说，被动句有三种译法：保留原语被动形式，借用汉语的"被""为""由""受到"等词语译出，如例（1）；或译成主动句，如例（2）；或译成无主句，如例（3）。例如：

(1) Earth's atmosphere is divided into three regions, namely troposphere, stratosphere and mesosphere.

译文：地球的大气层常被分成三层，分别命名为对流层、平流层和中间层。

(2) Furthermore, it is concerned with finding plausible solutions in the field of public health, such as waterborne diseases, implementing laws which promote adequate sanitation in urban, rural and recreational areas.

译文：此外，它还关注在公共卫生领域找到合理的解决办法，如水传播疾病，实施促进城市、农村和娱乐区充分卫生的法律。

(3) It was said that the developed countries should take the most of responsibility to

degradation of environment during the last 100 years.

译文：据说发达国家应该对过去 100 年的环境恶化承担最大责任。

3. 语 气

一般来说，英语中的语气有四种，即陈述语气、疑问语气、祈使语气和虚拟语气。从叙述方式上来说，科技英语常用陈述语气、祈使语气和虚拟语气。

祈使语气在很多专门用途英语，如商务英语、旅游英语、法律英语等中使用，这主要是由文体的性质决定的。科技英语文体中，祈使语气多用于产品说明书、操作规程、作业指导、程序建议以及注意事项等资料中。通常要求用精炼、简短的语言对产品进行介绍说明，以便用户对产品进行操作使用、对相关原理有所了解，强调语言叙述条理清晰、简洁、客观。使用祈使句，可使句子短小精悍。科技英语中，使用祈使句时，可直接以动词原形作为祈使句的开头，因为一般对需指明句子的主语而将其省略。例如：

(1) Do not touch the lamp inside or the lamp shade when the power is switched.

译文：台灯通电时，请不要触摸光源或灯罩。

无论是在日常英语，还是在正式文体英语中，虚拟语气都被广泛使用。科技英语中，虚拟语气主要用在以下从句中：条件从句（在非真实条件句中的应用）、状语从句（用在由 as if、as though 引导的方式状语从句中）、主语从句[通常为（should）be 句型，用于某些固定的句式中，如 It is necessary/ recommended that…等]。一般来说，科技英语文体中的虚拟语气多表示动作与事实相反，不可能实现或实现的可能性不大，其常用的标志词有 If、were、should 等。总的来说，在推测事物可能出现的变化趋势时，多用虚拟语气。例如：

(2) Were there no gravity, there would be no air around the earth.

译文：假如没有重力，地球周围就不会有空气。

(3) It is also necessary that the half-life and the beta energy (should) be determined for each isotope.

译文：还需测定每种同位素的半衰期和贝塔能。

翻译虚拟语气时，应注意其与真实条件句的区别。值得指出的是，虚拟语气中有时可省去连词 if，而将 should、were、had could 等提到主句前，采用倒装语序表达语气上的虚拟，如例（2）。对于表示建议、要求或命令等意义的虚拟句，其谓语结构形式为"should+动词原形"，有时动词采用被动形式，如例（3）。

二、科技英语的语篇特点

科技英语的语篇特点可以通过多个层面表现出来，如上述的词汇、句法、修辞等特点。但总体来说，科技英语呈现出来的语篇特征是语言规范、语气正式、陈述客观、逻辑性强、高准确性、信息量大、呈现高度专业化并大量使用公式、表格和图。

一般来说，科技英语的各种语篇特点不是单独呈现出来的，而是交叉出现的综合体。科技文体以描写自然现象、分析自然现象产生的规律、研究各种自然规律应用于人类生产实践的方法、表达所取得的各类成果及其运用等为主要内容,大致有专题著作、专题论文、教科书、实验报告、新闻报道、科普文章、仪器使用说明等表现形式。这些表现形式多为正式文体,用语规范,而且涉及各类专业领域,所以专业性极强。为了阐明所述事物发展的客观规律,必然要求语言运用者坚持客观公允的态度,力求准确地反映客观现实,因此需注重逻辑上的连贯,思维上的准确和严谨,表达上的清晰与精炼,以客观的风格陈述事实和揭示真理。例如:

The NCEP reanalysis project produced a retroactive 401 year record of global atmospheric fields and surface fluxes derived from a numerical weather prediction and data assimilation system kept unchanged over the analysis period. Use of a fixed model eliminates pseudo climate jumps in archived time series associated with frequent upgrades the operational modeling system used at NCEP and allows an assessment of the accuracy of a Numerical Weather Prediction (NWP) model over a long time period. However, temporal inconsistencies can still be present because of changes through time in the amount, type, and quality of the available assimilation data. The model used for the re-analysis is identical to the Medium-Range Forecast Model implemented operationally at NCEP in January 1995, except that the horizontal resolution is twice as coarse in the re-analysis version. Every 5 days, a single realization of an 8-day atmospheric forecast was run. For the period 1958—1998, this provides more than 2500 8-day forecasts that can become pared with observations.

译文:国家环境预报中心再分析项目整理了过去401年间全球大气场和地表通量的记录。在此分析时段,这些数据记录由同一数值天气预报和资料同化系统产生。使用固定模式可以消除在存档时期内因中心业务模式频繁升级而造成的一些伪气候突变,从而允许我们在较长时期内对数值天气预报的准确性进行评估。然而,因长时期内现有同化数据在数据总量、数据类型、数据质量方面的差异,时间的不一致性仍有可能出现。用来进行再分析的模式与中心1995年1月份起业务运行的中期天气预报模式相同,只不过再分析模式中每5天进行一次8天的天气预报,水平分辨率降低了一半。1958—1998年间,提供了超过2500次可以用来和实际观测进行对比的8日天气预报。

例文选自一篇气象学学术论文,不仅文体正式、语言规范,而且措辞严谨、专业性极强。文中涉及许多数据及观测(实验)设备,翻译中任何的理解或表达不当都会对其他参考该研究内容的科研工作者产生误导。

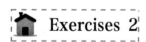

Exercises 2

1. Translate the following sentences into Chinese.

(1) In general, the faster the climate change, the greater will be the risk of damage. The mean sea level is expected to rise 9-88 cm by the year 2100, causing flooding of low lying areas and other damages.

(2) The ozone layer, therefore, is highly beneficial to plant and animal life on earth filtering the dangerous part of sun's radiation and allowing only the beneficial part to reach earth. Any disturbance or depletion of this layer would result in an increase of harmful radiation reaching the earth's surface leading to dangerous consequences.

2. Translate the following sentences into English.

（1）值得一提的是，在控制全球变暖的合作中，像中国这样的发展中国家一直在尽自己最大的努力，担负起自己的责任。

（2）假如地球的平均温度上升2℃，数以亿计的人会流离失所。

Part 3　Writing

科技报告类的书写

科技报告类的文体应简单、明了和高度概括。如果报告的内容超过1页，则从第2页起，在每页的上方标明页次、报告人姓名和报告日期，每个项目各占一行。科技报告的格式一般包括四个基本要素，即文头（heading）、简介（introduction）、讨论（discussion）和结论/建议（conclusion/recommendation），主要有实验报告（lab report）、进展报告（progressive report）、可行性报告（feasibility report）等。其中，实验报告是大学期间最常用的，下面我们就具体介绍一下实验报告的书写。

实验报告（experiment/ lab report）是对在实验室所做的实验的情况报告。实验报告主要包括以下四个方面的内容：①实验理由；②实验方法；③实验结果；④还要做哪些后续实验（如果必要）。

实验报告的文体和格式比较规范，一般的固定格式如下：

1 文头（heading）的内容与格式如下：

日期（date）

接收方（to）

报告人（from）

主题（subject）

2 简介（introduction）的内容有简述背景和目的，主要内容有：①实验理由；②实验目的；③权威性，也就是在谁的指导下完成的实验报告。

3 讨论（discussion）是实验报告的主体与实验方法，主要内容有：①实验设备和手段，包括实验仪器、方法以及理论依据；②实验步骤，要按时间顺序进行编写。

4 结论/建议（conclusion/ recommendation）包括两个方面的内容：①结论。要在实验报告中提供实验中的新发现，如何解释这些发现，这些发现的意义所在。②建议是指对在下一步的实验中有必要采取的措施。

此外，也可以使用图表对实验报告进行充实。在实验报告中使用略图和线路图来说明实验过程。如：例1是一个实验报告的实例。

【例1】实验报告（lab report）

Date: July 18, 2019
To: Dr. Jones
From: Sam Ascendio, Lab Technician
Subject: LAB REPORT ON THE ACCURACY OF DECIBEL VOLTAGE GAIN （A） MEASUREMENTS

INTRODUCTION

Purpose

Some apparatus has been found inaccuracies in recent measurements. In response to their request, this report will present the result of tested A (gain in decibels) of our ABC voltage divides circuit. Measured A will be compared to calculated A. This will determine the accuracy of the measuring device.

DISCUSSION

Apparatus

- Audio generator
- Decade resistance box
- 1/2 W resistor, four 470 ohm, two 1 kilohm, 100 kilohm
- AC millivoltmeter

Procedure

1. With a voltage divider, for each value of R (resistances) in Table 1, voltage gain was calculated (table attached).

2. An audio generator was adjusted to give a reading of o dB for input voltage.

3. Output voltage was measured on the dB scale. This reading the measured A and is recorded in Table 1.

4. Step 3 was repeated for each value of R listed in Table 1

5. $A_1=V_1/V_2$, $A_2=V_2/V_3$, and $A_3=V_3/V_4$. These voltage gains added together give the total voltage, as recorded in Table 2

6. The circuit in voltage divider was connected. Input voltage was set at o dB on the 1-V range of the AC millivoltmeter.

7. Values V_2, V_3, and V_4 were read and recorded in Table 3.

TABLE 1

R (kilohm)	Calculated A (dB)	Measured A (dB)
240	−3.02	−3.21
100	−6.02	−6.13
46	−10.00	−9.71

| 11 | −20.00 | −20.00 |
| 1 | −40.00 | −40.00 |

TABLE 2		TABLE 3	
A_1	0	V_1	0 dB
A_2	0.50	V_2	−5.97 dB
A_3	0.66	V_3	−11.97 dB
		V_4	−18.06 dB

CONCLUSION/ RECOMMENDATION

Accuracy between measured and calculated in Table 1 was between 0.1 and 0.3 dB. This is acceptable. Accuracy between calculated A in Table 2 and calculated voltage gain in Table 3 was 0.1 dB, also very accurate. These tests show that no further action should be taken.

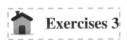 Exercises 3

Write a lab report with no less than 200 words, based on the requirements of scientific English Writing.

 Expanding Reading

EPA Science Matters Newsletter: Impacts of Climate Change

Climate change is happening. Global average temperature has increased about 1.8°F from 1901 to 2016. Changes of one or two degrees in the average temperature of the planet can cause potentially dangerous shifts in climate and weather. For example, many places have experienced changes in rainfall, resulting in more floods, droughts, or intense rain, as well as more frequent and severe heat waves.

The planet's oceans and glaciers have also experienced changes—oceans are warming and becoming more acidic, ice caps are melting, and sea level is rising. As these and other changes become more pronounced in the coming decades, they will likely present challenges to our society and our environment.

Climate change impacts our health, environment, and economy. For example:
- Warmer temperatures increase the frequency, intensity, and duration of heat waves, which can pose health risks, particularly for young children and the elderly.
- Climate change can also impact human health by worsening air and water quality, increasing the spread of certain diseases, and altering the frequency or intensity

of extreme weather events.
- Rising sea level threatens coastal communities and ecosystems.
- Changes in the patterns and amount of rainfall, as well as changes in the timing and amount of stream flow, can affect water supplies and water quality and the production of hydroelectricity.
- Changing ecosystems influence geographic ranges of many plant and animal species and the timing of their lifecycle events, such as migration and reproduction.
- Increases in the frequency and intensity of extreme weather events, such as heat waves, droughts, and floods, can increase losses to property, cause costly disruptions to society, and reduce the affordability of insurance.

(*https://www.epa.gov/climatechange-science/impacts-climate-change*)

Unit 2 Air Pollution and Control

Lesson 1 Air Pollution

Part 1 Reading

Air emissions have natural or anthropogenic origins; however, this book focuses primarily on the latter, which include industrial activities and the burning of fossil fuels. These constituents of pollution have the potential to affect the majority of people in a region.

Air pollutants comprise primary and secondary air pollutants. Primary air pollutants are emitted directly from sources. They include, but are not limited to, particulate matter (PM), sulfur dioxide (SO_2), nitric oxides (NO_x), hydrocarbon (HC), volatile organic compounds (VOCs), carbon monoxide (CO), and ammonia (NH_3). Secondary air pollutants are produced by the chemical reactions of two or more primary pollutants or by reactions with normal atmospheric constituents. Examples of secondary air pollutants are ground level ozone, formaldehyde, smog, and acid mist.

Particulate matter is a mixture of solid particles and liquid droplets suspended in the air. Particulate matter is interchangeable with aerosol, which is a suspension of solid or liquid particles in a gas. It is a two-phase system consisting of particles and the gas in which they are suspended. Particulate matter can be both primary pollutants and secondary pollutants that are sent directly into the atmosphere in the form of windblown dust and soil, sea salt spray, pollen, and spores[①]. Other examples of PM are smoke, fumes, and haze.

For particulate matter where particle diameters are smaller than x micrometers, it is defined as PM_x. Commonly used terms are PM_{10} and $PM_{2.5}$. Sometimes, particulate matter and aerosol is exchangeable. Monodisperse aerosols, in which all particles have the same size, can be produced in laboratory for use as test aerosols. In practice, engineers deal with polydisperse aerosols (i.e. suspended particles are in a wide range of sizes), and statistical measures should be used to characterize particle sizes[②]. Aerosol Technology by William Hinds (2006) is one of the reference books for this subject.

Air pollutants other than PM present primarily as gases. Volatile organic compounds (VOCs) are chemicals that contain carbon and/or hydrogen and evaporate easily. VOCs are the main air emissions from the oil and gas industry, as well as indoor consumer products and construction materials, such as new fabrics, wood, and paints. VOCs have been found to be a major contributing factor to ground-level ozone, a common air pollutant, and a proven public health hazard. Sulfur dioxide (SO_2) and nitric oxides (NO_x) are two major gaseous air pollutants generated through combustion processes. Carbon monoxide (CO) and hydrocarbon (HC) are generated from incomplete combustion and are converted into CO_2 through a complete combustion process.

Secondary air pollutants are those formed through complex physical and/or chemical reactions.Air pollution is an evolving subject and inevitable, as the demand for energy increases. Air pollution really flourished with the Industrial Revolution and continues to grow with the human appetite for comfort and speed.

At first, the study of air pollution focused on recurring episodes of high levels of air pollution in areas surrounding industrial facilities, such as coal burning power plants and chemical refineries. These pollution episodes were accompanied by acute human sickness and the exacerbation of chronic illness. After the mid-twentieth century, when industrialized nations' economies recovered rapidly from World War II, many urban regions without heavy industrial facilities also began to experience high levels of photochemical smog and nitrogen oxides[3].

The twentieth century marked the beginning of the understanding that human activity was having deleterious effects upon the natural world, including human health and welfare. These effects included increasing pollution of air, water, and land by the byproducts of industrial activity, and the permanent loss of natural species of plants and animals through changes in laboratory settings, water usage, and human predation.

The topic of indoor air quality (IAQ) has become popular, due to the awareness of asthma and allergies triggered by indoor air pollutants such as mold. IAQ awareness also increased with the involvement of the United States Environmental Protection Agency. The energy crisis in the 1970s resulted in tighter building envelope, sealing, and insufficient ventilation[4]. Most existing heating, ventilation, and air conditioning (HVAC) systems were designed for temperature control without consideration of air pollutant accumulations. As a result, IAQ degraded, and problems arose. Recent findings have demonstrated that indoor air is often more polluted than outdoor air in many developed countries, thereby causing a greater health concern as current lifestyles demand more time indoors.

The later twentieth and early twenty-first centuries saw a boom in nanotechnology. Nanotechnology has been tested for air quality remediation in such areas as noncatalytic combustion and photocatalytic oxidation of volatile organic compounds (VOCs). On the other hand, the environmental effects of nanotechnology are not well understood; and, concerns have recently begun to increase. The world is not ready for nanotechnology because "the future is coming sooner than it is expected". The effect of nanotechnology to air quality is still waiting for systematic studies to confirm its environmental effects. Scientific evidence is needed before definitive conclusions can be made.

Words and Expressions

anthropogenic	[ænθrəpəʊˈdʒɛnɪk]	adj. 人为的
fossil fuels		化石燃料
hydrocarbon	[ˌhaɪdrəˈkɑːbən]	n. 烃、碳氢化合物
primary pollutant		一次污染物
secondary pollutants		二次污染物
reaction	[rɪˈækʃn]	n. 反应、感应
formaldehyde	[fɔːˈmældɪhaɪd]	n. 甲醛，福尔马林
interchangeable	[ˌɪntəˈtʃeɪndʒəbl]	adj. 可交换的，可互换的，可交替的
aerosol	[ˈeərəsɒl]	n. 气溶胶、气雾剂、（喷油漆、头发定型剂等的）喷雾器、雾化器
spore	[spɔː(r)]	n. 孢子
suspended particulate matter （SPM）		悬浮颗粒物
monodisperse		adj. 单分散（性）的
polydisperse		adj. 多分散（性）的
volatile organic compounds (VOCs)		挥发性有机化合物
appetite	[ˈæpɪtaɪt]	n. 食欲，胃口，强烈欲望
episode	[ˈepɪsəʊd]	n. 插曲，（人生的）一段经历
photochemical smog		光化学烟雾
exacerbation	[ɛksˌæsə(ː)ˈbeɪʃən]	n. 加重，恶化，激怒
chronic	[ˈkrɒnɪk]	adj. 慢性的，长期的
deleterious	[ˌdeləˈtɪərɪəs]	adj. 有害的，造成伤害的，损害的
asthma	[ˈæsmə]	n. 气喘、哮喘
allergy	[ˈælədʒi]	n. 过敏，变态反应
mold	[məʊld]	n. 模具，模型，压模
ventilation	[ˌventɪˈleɪʃən]	n. 通风，通气能力，通气量，通风设备
accumulation	[əˌkjuːmjəˈleɪʃn]	n. 积累，堆积物，聚积物
nanotechnology	[nænəʊtekˈnɒlədʒi]	n. 纳米技术
remediation	[rɪˌmiːdiˈeɪʃn]	n. 补救
noncatalytic		adj. 非催化的
photocatalytic		adj. 光催化的
definitive	[dɪˈfɪnətɪv]	adj. 最后的，决定性的

Notes

① Particulate matter can be both primary pollutants and secondary pollutants that are sent directly into the atmosphere in the form of windblown dust and soil, sea salt spray, pollen, and spores.

参考译文：颗粒物既可以是一次污染物，也可以是二次污染物，它以灰尘、土壤、海盐雾、花粉和孢子等形式被风直接吹入大气。

② In practice, engineers deal with polydisperse aerosols (i.e. suspended particles are in a wide range of sizes), and statistical measures should be used to characterize particle sizes.

参考译文：在实践中，工程师要处理多分散气溶胶（即悬浮颗粒的尺寸范围很广），应使用统计方法来表征颗粒尺寸。

③ After the mid-twentieth century, when industrialized nations' economies recovered rapidly from World War Ⅱ, many urban regions without heavy industrial facilities also began to experience high levels of photochemical smog and nitrogen oxides.

参考译文：20世纪中叶之后，工业化国家的经济从第二次世界大战中迅速复苏，许多没有重工业设施的城市地区也开始遭遇高浓度的光化学烟雾和氮氧化物。

④ The energy crisis in the 1970s resulted in tighter building envelope, sealing, and insufficient ventilation.

参考译文：20世纪70年代的能源危机导致建筑围护结构更加紧密、密封性更好，同时通风不足。

Exercises 1

1. According to the reading material, chose the best answer(s) from the options.

(1) Air pollutants comprise _____ and _____ air pollutants

A. primary　　　B. secondary　　　C. transform　　　D. disappearance

(2) Which of the following is NOT showed as air pollutants in the article?

A. SS　　　B. CO　　　C. NO_x　　　D. SPM

(3) For particulate matter where particle diameters are _____ than x micrometers, it is defined as PM_x.

A. heavier　　　B. bigger　　　C. smaller　　　D. lighter

(4) Human activity was having deleterious effects upon the world. These effects included _____ , _____ , and _____ , and the permanent loss of natural species of plants and animals through changes in laboratory settings.

A. increasing pollution of air　　　B. human predations

C. asthma　　　D. water usage

(5) _____ has been tested for air quality remediation in such areas as noncatalytic combustion and photocatalytic oxidation of volatile organic compounds (VOCs).

A. IAQ B. Nanotechnology C. HVAC D. Photocatalytic

2. Propose the generation of 3 kinds of air pollutants at least.

3. Establish a timeline of the researches in different decades.

Part 2　Translation

英语翻译的标准

翻译标准，就是指翻译实践中译者所依据的准则。切实可行的翻译标准对发挥翻译功能、提高翻译质量具有重要的意义。古今中外，翻译实践者和学者都提出过不少翻译标准。其中，影响最大的当属严复的"信、达、雅"三字标准。

清末的严复于1897年在《天演论》序言中提出了"信、达、雅"之说。所谓"信"，就是译文的思想内容忠实于原作；"达"就是译文表达通顺明白；"雅"即要求风格优美。

后来，鲁迅在《且介亭杂文二集》提出了"信顺"之说翻译标准；茅盾在《为发展文学翻译事业和提高翻译质量而奋斗》中提出了"文学的翻译是用另一种语言，把原作的艺术意境传达出来，使读者在读译文的时候能够像读原作时一样得到启发、感动和美的感受"；傅雷在《高老头》重译本序中提出了"神似"之说；钱钟书提出了"化境"之说；时至今日，人们还在提出新的翻译标准。其实，每个翻译标准都是在一定的时代或环境中提出来的，没有一个"放之四海皆准"的翻译标准。

为了避免歧义，一般的翻译还是应当遵守"忠实、通顺"四字标准的。就科技翻译而言，除了以上"忠实、通顺"标准外，还应遵守"规范"标准。

一、忠　实

忠实是指译文忠实于原作的内容。译者必须把原作的内容完整而准确地表达出来，使译文读者得到的信息与原文读者得到的信息大致相同，不能有歪曲、遗漏或其他的删改等。例如：

(1) "Site" refers to the land and other places on, under, in or through which the Permanent Works or Temporary Works designed by the Engineer are to be executed.

译文1："现场"指工程师设计的永久性或临时性工程所需的土地及其他场地。

译文2："现场"指工程师设计的永久工程或临时工程所需的土地和其他场所，包括地面、地下、工程范围之内或途经的部分。

对比译文和原文，不难发现，译文1在翻译过程中漏译了介词on、under、in、through

33

所涵盖的意义，最后导致译文不够忠实于原文，违背了翻译所要求的忠实原则，译文2则忠实于原文且通顺得体。

另外，值得注意的一个问题就是形式的忠实与内容的忠实。文字的内容和形式是统一的关系。内容是思想，形式是表达这一思想所用的语言、词汇、词组和句子结构。内容是原文的灵魂，形式是原文的外壳，它们不是一个东西，但又密不可分。如果对原文理解不够，不能透过原文的形式掌握其精神实质，翻译的时候就有可能过多地受原文形式的拘束，迁就于原文的字面和结构。这样翻译出来的文字，形式上与原文似乎一致了，但与原文的意思却相去甚远。例如：

(2) They are good questions, because they call for thought-provoking answers.

译文1：它们是精彩的问题，因为它们需要对方做出激发思想火花的回答。

译文2：这些问题问得好，要回答好就要好好动一下脑筋。

译文1中的两个"它们"亦步亦趋，有点不伦不类，读起来不像地道的中文，是字字对应的形式忠实，而"激发思想火花"的说法过于生硬，远不如"好好动一下脑筋"来得通俗易懂，虽然表达形式有所变化，但内容却更加忠实。

二、通　顺

通顺是指译文的语言必须通俗易懂，符合汉语的表达规范。换言之，译文的表达要按汉语的语法和习惯来选词造句，没有文理不通、结构混乱、逻辑不清的现象。钱歌川在《翻译的基本知识》（1981）中举例说，辜鸿铭曾以"汉滨读易者"的笔名，著有《张文襄幕府纪闻》一书，其中有这么一则故事："昔年陈立秋侍郎兰彬，出使美国，有随员徐某，夙不解西文。一日，持西报展览，颇入神。使馆译员见之，讶然曰：君何时谙识西文乎？徐曰：我固不谙。译员曰：君既不谙西文，阅此奚为？徐答：余以为阅西文固不解，阅诸君之翻译文亦不解。同一不解，固不如阅西文之为愈也。至今传为笑柄。"由上面这个故事看来，翻译出来的文字必须通顺达意，否则就会变成天书，是没有人能看得懂的。无怪乎徐先生那样不懂英文的人，宁愿放下翻译好的本国文字，却入神地去看那天书似的原文了。他的幽默感给了从事翻译的人们当头一棒，使他们在执笔翻译之前，先得想想：他们是翻译谁看的？当然是翻给他们的国人看的。如果国人看不懂，岂不等于劳力白费？鲁迅先生说翻译必须"力求其易解"，也就是这个意思。这也是通常情况下译者所必须遵守的准则。例如：

(1) Action is equal to reaction, but it acts in a contrary direction.

译文1：作用力与反作用力相等，但它向着相反的方向起作用。

译文2：作用力和反作用力大小相等，方向相反。

应该说，译文1和原文的意思基本吻合，没有产生误译。由于原文属科技文体，译者虽然能基本传达原文的基本意思，但是却没有了科技味，也就是说，译译者只是"钻进去"而没有"跳出来"，译文不够简练通顺，违背了翻译的应有之义——通顺。而译文2则言简意赅，不落窠臼，完全摆脱了原文形式的束缚，选用了由四字词构成的对偶修辞手法，使译文显得格外生动贴切。

(2) An atom consists of the nucleus and one or more electrons moving around it.

译文 1：一个原子由原子核和一个或更多个绕着原子核运动的电子所组成。

译文 2：原子由原子核和一个以上绕核运动的电子组成。

译文 1 虽然表述正确，但较为啰嗦，仅"原子"一词就重复数次，造成译文表达不够通顺。译文 2 则简化了译文表述，通俗易懂。

关于忠实与通顺的关系，很多人吐槽翻译过程中，忠实和通顺是很难两全的。事实上，绝大多数的情况并非如此，忠实和通顺不但可以两全，而且还能合二为一。例如：

(3) Newton brought together the discoveries of Copernicus, Kepler, Galileo, and others in astronomy and in dynamics. To these he added his own findings and fused them into a structure that still stands today, one of the greatest achievements of science.

译文：牛顿将哥白尼、开普勒、伽利略等人以及自己在天文学和力学方面的发现综合汇集，揉为一体。时至今日，他的这一成果仍不失为最伟大的科学成就之一。

译文打破了原文的结构，对原文的表达形式进行了巧妙的、合理的、大幅度的变通，不仅忠实于原文，而且表达流畅自如、别开生面，是忠实与通顺的佳句。

总之，忠实是通顺的基础，通顺是忠实的保证。不忠实于原文而片面追求译文的通顺，则译文就失去自身的价值，成为无源之水，无本之木。但是，不通顺的译文，使读者感到别扭，也必然影响对原文的准确表达，也就谈不上忠实了。可见，忠实与通顺是对立的统一，两者的关系反映了内容与形式的一致性。忠实和通顺定要两者兼顾，把握好分寸，相映成趣。

三、规 范

所谓规范，就是说科技翻译中，译文不仅要忠实于原文，表达流畅，而且必须符合科技英语的表达规范，即专业、规范。由于科技英语专业性较强、术语较多，故翻译过程中，译文的专业术语表述要符合科技语言和术语的规范，尽可能利用译入语中已有的约定俗成的定义、术语或概念。例如在汽车制造业中， side valve 不是指"侧面阀"，而是指"侧置气门"，因为 vale 作为汽车零件不是通常意义上的"阀门"，而是指发动机的"气门"。如果把 one of the valves in the engine must have gone wrong 译为"发动机里的一个阀门肯定出了问题"，则会贻笑大方。总之，科技翻译中，专业、规范是极为重要的。例如：

(1) Traditionally, rural highway location practice has been field oriented, but the modern method is "office" oriented.

译文：传统上，乡村公路定线采用现场定线法，而现在的方法则是采用纸上定线或计算机定线。

原文中的"field oriented"和"office oriented"都是行业用语，极具专业性。如果将其译为"田野导向"和"办公室导向"，而不是"现场定线法"和"纸上或计算机定线"，则失去了其专业、规范的含义。

(2) The Red Cross has begun a major cloning project relating to the production of transgenic pigs for organ donors.

译文1：红十字会已开始一项重大的克隆工程，该工程与培育用于器官捐献的转基因的猪有关。

译文2：红十字会已开始一项重大的克隆工程，培育做器官供体之用的转基因猪。

译文1中将 transgenic pigs for organ donors 译为"用于器官捐献的转基因的猪"，颇令人费解，原因在于译者没有完全理解其中的逻辑含义，又不熟悉专业术语。从语言关系上来说，培养转基因的猪（production of transgenic pigs）是为了让猪成为器官捐献者（for organ donors）。从专业上来讲，organ donors 即为"器官供体"。译文2则既忠实、通顺，又规范、得体。

 **Exercises 2**

Translate the following paragraph into Chinese.

The later twentieth and early twenty-first centuries saw a boom in nanotechnology. Nanotechnology has been tested for air quality remediation in such areas as noncatalytic combustion and photocatalytic oxidation of volatile organic compounds (VOCs). On the other hand, the environmental effects of nanotechnology are not well understood; and, concerns have recently begun to increase. The world is not ready for nanotechnology because "the future is coming sooner than it is expected". The effect of nanotechnology to air quality is still waiting for systematic studies to confirm its environmental effects. Scientific evidence is needed before definitive conclusions can be made.

Part 3　Writing

学术论文的撰写（一）——学术论文的特点

学术论文是作者对自己研究成果的展示，或对未来科学发展的展望，它在语言、内容及结构上不同于其他文体。

一、语言上的特点

1. 用词规范

就用词而言，学术论文应遵循三个原则：

（1）必须用技术词语，以使论文更专业、更科学、更有学术性。

（2）词义必须客观、准确，不能华而不实。

（3）词或短语必须是正式的书面用语。

以下短语往往不用在学术论文中：

Here, I'd like to have a talk about…

Well, I think…

It's…

He's…

Isn't…

2. 表达完整、严谨

就表达而言，学术论文应满足四个要求：

（1）要用完整的句子来表达，且要多用复合句和长句子，使其结构严谨。

（2）一般常用的时态为一般现在时、现在完成时和一般将来时，在可能的情况下，被动语态可用得更多些。

（3）句与句之间要通过有效、适当的连接手段，使全文结构合理，层次分明。

（4）论证推理要合情合理，逻辑性要强。

二、结构上的特点

1. 有固定的格式

由于学术论文反映了一个国家科学技术的水平，中国也以国标的形式（GB 7713—87）规定了中国学术论文写作的格式，如图2.1所示。

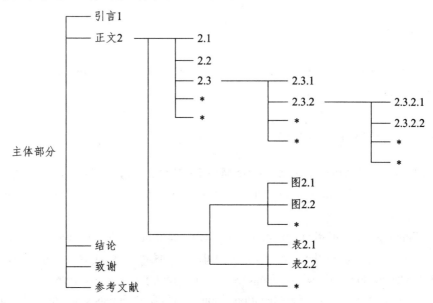

图 2.1　中国学术论文写作格式

2. 有通用的编码

（1）中图分类号

中图分类号，是指采用《中国图书馆分类法》（简称《中图法》）对科技文献进行主题分析，并依照文献内容的学科属性和特征，分门别类地组织文献，所获取的分类代号。

《中图法》是中华人民共和国成立后编制出版的一部具有代表性的大型综合性分类法，是当今国内图书馆使用最广泛的分类法体系。包括马列主义、毛泽东思想，哲学，社会科学，自然科学，综合性图书五大部类，22个基本大类。采用英文字母与阿拉伯数字相结合的混合号码，用一个字母代表一个大类，以字母顺序反映大类的次序，在字母后用数字做标记。为适应工业技术发展及该类文献的分类，对工业技术二级类目，采用双字母。

学术论文的中图分类号可以方便地通过互联网查询。

（2）文献标识码

每篇文章或资料应有一个文献标识码，并且规定了与每种文献标识码相对应的文献中的数据项，即格式。标识码（Document code）是按照《中国学术期刊（光盘版）检索与评价数据规范》规定的分类码，作用在于对文章按其内容进行归类，以便于文献的统计、期刊评价、确定文献的检索范围，提高检索结果的适用性等。具体如下：

A——理论与应用研究学术论文（包括综述报告）；
B——实用性技术成果报告（科技）、理论学习与社会实践总结（社科）；
C——业务指导与技术管理性文章（包括领导讲话、特约评论等）；
D——一般动态性信息（通信、报道、会议活动、专访等）；
E——文件、资料（包括历史资料、统计资料、机构、人物、书刊、知识介绍等）。

Exercises 3

Draw a tree diagram of the structural characteristics of one academic paper published in journal of environmental protection field.

Moving CO$_2$ from Air to Oceans May Be Necessary to Slow Warming

A major new report from the National Academies examines options to store carbon in the oceans

Climbing concentrations of carbon dioxide make it likely that humans will have to move some gases from the atmosphere into the oceans to prevent crippling effects of climate change, the National Academies said in a major report released yesterday.

It came after months of deliberation among top U.S. scientists who concluded that global efforts to reduce emissions, even if successful, "may not be enough to stabilize the climate." The report identified six ways to capture and store carbon dioxide in the oceans, a controversial idea that the report said "will likely be needed."

Potential methods include stimulating more plant growth in and around the oceans and manipulating ocean currents to draw CO$_2$ deep underwater.

The report by the National Academies of Sciences, Engineering and Medicine (NASEM) is cautiously written and points out that research for ocean-based CO_2 reductions has only recently begun. It noted that "showstoppers" could be ahead, including prohibitive costs and legal conflicts.

On the other hand, it pointed out that oceans cover 70 percent of Earth, and they already remove and store a "substantial fraction" of the CO_2 being released through burning fossil fuels and other human activities.

Scott Doney, the chair of the committee that prepared the report, said during a press briefing yesterday that there is "increasing interest from businesses and entrepreneurs" in finding ocean-based solutions to climate change.

"The scientific community needs to step up and help," he said.

The nonprofit Aspen Institute responded to the National Academies report with its own study yesterday calling for an international "code of conduct" to govern sea-based carbon removal experiments.

"We are excited to see such a rigorous look at ocean-based carbon dioxide removal research," said Greg Gershuny, executive director of the Aspen Institute's energy and environment program. "We need to explore all viable, creative climate solutions, but we also need to ensure there are good governance and environmental justice considerations rooted in the technologies we develop," he said.

The National Academies offered a menu of potential experiments, beginning with one that has been studied since the 1950s. It may be among the simplest and the cheapest approaches offered in the report.

It's called "nutrient fertilization" and most of its experiments up to now have involved scattering iron filings over selected ocean areas to stimulate the growth of microscopic plants called phytoplankton.

The plants absorb CO_2 in sunlight and are eaten by tiny organisms called zooplankton. In the process, the carbon gradually sinks with the plankton, and the fish that eat it, to deeper ocean waters where the CO_2 is believed to remain for at least a century.

A second approach could help stimulate downward motion of the plankton by using wave power or by injecting cooler water. But there are "gaps in understanding" about how that might be scaled up, the report said.

A third method involves the large-scale farming of seaweed in the ocean. It's already underway to produce food for people and animals. It's also being studied for the production of low carbon biofuels.

Seaweed stores the carbon made from sunlight, and it might then be pumped to lower depths where the CO_2 can be sequestered. Again, there are "gaps" in how this might work, the report said.

Oceans store "more than 50 times as much carbon as the atmosphere does," the report says, so other ways of stimulating storage of CO_2 are being considered. They include lowering the acid content of oceans by dumping powder derived from grinding rocks into the sea.

Then there is something called "electrochemistry," which can be used to generate electrolysis to remove CO_2 from seawater and store it in rock formations on land.

Another approach is to encourage the formation of coastal forest systems such as mangroves, salt marshes and sea grasses, which, like inland forests, remove CO_2 from the air.

Researchers will start with laboratory experiments and then pick the successful ones for "pilot field-scale" experiments, the National Academies report says, adding that "adequate environmental and social risk reduction measures" will be built into the process.

This article was originally written By John Fialka, published on E&E News on December 9, 2021

（https://www.scientificamerican.com/article/moving-co2-from-air-to-oceans-may-be-necessary-to-slow-warming/）

Unit 2 Air Pollution and Control

Lesson 2   Air Pollution Emission Control Devices for Stationary Sources

Part 1   Reading

1 Introduction

Stationary sources of air pollution emissions, such as power plants, steel mills, smelters, cement plants, refineries, and other industrial processes, release contaminants into the atmosphere as particulates, aerosols, vapors, or gases. These emissions are typically controlled to high efficiencies using a wide range of air pollution control devices. The selection of the appropriate control technology is determined by the pollutant collected, the stationary source conditions, and the control efficiency required. In some cases, pollutant emissions can be reduced significantly through process modifications and combustion controls. However, in most instances, some form of add-on pollution control equipment is installed in the ductwork (or flues) leading to the smoke stack to meet current allowable emission limits.

Common methods for eliminating or reducing gaseous pollutants include:

• Destroying pollutants by thermal or catalytic combustion, such as by use of a flare stack, a high temperature incinerator, or a catalytic combustion reactor.

• Changing pollutants to less harmful forms through chemical reactions, such as converting nitrogen oxides (NO_x) to nitrogen and water through the addition of ammonia to the flue gas in front of a selective catalytic reactor.

• Collecting pollutants using air pollution control systems before they reach the atmosphere.

The most commonly used devices for controlling particulate emissions include:

• Electrostatic precipitators (wet and dry types)
• Fabric filters (also called bag houses)
• Wet scrubbers
• Cyclones (or multiclones)

In many cases, more than one of these devices is used in series to obtain desired removal efficiencies for the contaminants of concern. For example, a cyclone may be used to remove large particles before a pollutant stream enters a wet scrubber.

Common control devices for gaseous and vapor pollutants include:
- Thermal oxidizers
- Catalytic reactors
- Carbon adsorbers
- Absorption towers
- Biofilters

2 Absorption &Wet Scrubbing Equipment

Scrubbing is a physical process whereby particulates, vapors, and gases are controlled by either passing a gas stream through a liquid solution or spraying a liquid into a gas stream. Water is the most commonly used absorbent liquid. As the gas stream contacts the liquid, the liquid absorbs the pollutants, in much the same way that rain droplets wash away strong odors on hot summer days.

Gas absorption is commonly used to recover products or to purify gas streams that have high concentrations of water-soluble compounds. Absorption equipment is designed to get as much mixing between the gas and liquid as possible.

Common types of gas absorption equipment include spray towers, packed towers, tray towers, and spray chambers. Packed towers are by far the most commonly used control equipment for the absorption of gaseous pollutants. However, when used with heavy, particulate-laden gas, they can be plugged by particulate matter (PM). Wet collection devices used for PM control include venturi scrubbers, bubbling scrubbers, spray towers, and in some instances, wet electrostatic precipitators (ESPs).

3 Adsorption

The process of adsorption involves the molecular attraction of gases or vapors [usually volatile organic compounds (VOCs)] onto the surface of certain solids (usually carbon, molecular sieves, and/or catalysts). This attraction may be chemical or physical in nature and is predominantly a surface effect. Activated carbon (charcoal), which possesses the large internal surface area needed to adsorb large quantities of gases within its structure, is often used to remove VOCs from flue gases. After the activated carbon is saturated with VOCs, it is often treated (by heat and/or steam) to strip off the collected VOCs. The VOCs are then sent for further treatment, and the carbon is reused in the adsorption reactor. Adsorption is affected by the temperature, flowrate, concentration, and molecular structure of the gas.

Adsorption is commonly used for removing gases from contaminated soil, oil refineries, municipal wastewater treatment plants, industrial paint shops, and steel mills.

4 Fabric Filters or Bag Houses

Fabric filters, also commonly referred to as bag houses, are used in many industrial applications. They operate in a manner similar to a household vacuum cleaner. Dust-laden gases pass through fabric bags where the dry particulates are captured on the fabric surface. After enough dust has built up on the filters, as indicated by a build up in pressure across the fabric, dust is periodically removed by blowing air back through the fabric, pulsing the fabric with a blast of air, or shaking the fabric[①]. Dust from the fabric then falls to a collection hopper where it is removed. As dust builds up on the fabric, the dust layer itself can act as a filter aid improving the removal efficiency of the device.

Bag houses are used to control air pollutants from coal-fired power plants, steel mills, foundries, and other industrial processes. Fabric filers can collect over 99.9% of the entering particulates, even fine PM. Bag houses also are sometimes used as part of a multistage gas cleaning system where they are used as a reactor as well as a particulate removal device, such as in semi-dry flue gas desulphurization systems. Recently, some bag houses are being equipped with catalytic bags where they also act as a chemical reactor while they are collecting particulate.

5 Catalytic Reactors

Catalytic reactors, referred to as selective catalytic reduction (SCR) systems, are use extensively to control NO_x, emissions arising from the burning of fossil fuels in industrial processes. Ammonia is injected and mixed with the flue gases upstream of the SCR reactor. In the SCR reactor, ammonia and NO_x, react to form nitrogen and water. Greater than 90% NO_x, removal is possible with these systems.

Catalytic reactors also perform thermal destruction functions like incinerators, but at lower temperatures and for selected waste gases only. They incorporate beds of solid catalytic material that the unwanted gases pass through, typically for oxidation or reduction purposes. Catalytic reactors have the advantage of lowering the thermal energy requirements. Destruction efficiencies of 99.99% are possible with significantly reduced energy and operating costs as compared to an incinerator.

A very common example of a catalytic reactor is the three-way catalytic converter used in modern automobiles to simultaneously reduce emissions of NO_x, VOCs, and carbon monoxide.

6 Cyclones

Dust-laden gas is whirled rapidly inside a collector shaped like a cylinder (or cyclone). The swirling motion creates centrifugal forces that cause the particles to be thrown against the walls of the cylinder and drop into a hopper below. The gas left in the middle of the cylinder after the dust particles have been removed moves upward and exits the cylinder. Cyclones operate to collect relatively large size PM from a gaseous stream, and can operate at elevated temperatures. Cyclones are typically used for the removal of particles 50 microns (μm) or larger. Efficiencies greater than 90% for particle sizes of 10 μm or greater are possible, and efficiency increases exponentially with particle diameter and with increased pressure drop through the cyclone.

Cyclones are widely used; they control pollutants from cotton gins, rock crushers, and many other industrial processes that contain relatively large particulate in the gas stream. They can be used to remove either solid particles or liquid droplets, Cyclones can experience a number of problems including particles recirculating from the hopper, and erosion and corrosion of the cyclone internals due to the nature of the material being collected （corrosive and/or abrasive）. Heavy dust at the inlet of the cyclone can also lead to plugging of the cyclone hopper.

7　Electrostatic Precipitators (ESPs)

ESPs are relatively large, low velocity dust collection devices that remove particles in much the same way that static electricity in clothing picks up small pieces of lint. Transformers are used to develop extremely high voltage drops between charging electrodes and collecting plates. The electrical field produced in the gas stream as it passes through the high voltage discharge introduces a charge on the particles, which is then attracted to the collecting plates[②]. Periodically the collected dust is removed from the collecting plates by a hammer device striking the top of the plates (rapping) dislodging the particulate, which falls to a bottom hopper for removal.

Electrostatic precipitators are often configured as a series of collecting plates to improve overall collection efficiency. Efficiencies exceeding 99% can be achieved, and ESPs are used in many of the same applications as bag houses, including power plants, steel and paper mills, smelters, cement plants, and petroleum refineries. In some applications water is used to remove the collected particulates. ESPs using this cleaning mechanism are referred to as "wet ESPs" and are often used to remove fumes such as sulfuric acid mist.

8　Incinerators

Incineration involves the high efficiency combustion of certain solid, liquid, or gaseous wastes. The reactions may be self-sustaining based on the combustibility of the waste, or may require the addition of auxiliary fuels, such as natural gas or propane. They may be batch operations or continuous as with flares used to burn off methane from landfills. When not burning solids, they are also called thermal oxidizers, and these devices can operate at efficiencies of 99.99% (as with hazardous waste incinerators).

Thermal oxidizers are used to destroy odorous or toxic VOCs. Achieving the required temperatures (up to 2 000℉) requires a large fuel usage, and costs can be high. Regenerative thermal oxidizers, though, can achieve very high heat recoveries (up to 95%), greatly reducing fuel costs.

9 Biofilters

Biofilters operate to destroy VOCs and odors by microbial oxidation of these problem compounds. They are most effective on water-soluble materials. The polluted air is passed through a wetted bed, which supports a biomass of bacteria that absorb and metabolize pollutants. Efficiencies over 98% are possible with this application.

(*http://www.events.awma.org/files_original/ControlDevicesFact sheet07.pdf*)

Words and Expressions

spray tower		喷淋塔
packed tower		填料塔
venture scrubber		文丘里洗涤器
bubbling scrubber		鼓泡洗涤器
volatile organic compound		挥发性有机物
activated carbon		活性炭
charcoal	['tʃɑːkəʊl]	n. 炭，木炭
saturate	['sætʃreɪt]	v. 使充满，使饱和
strip off		脱去
flowrate	['fləureit]	n. 流量，流速
vacuum	['vækjuːm]	n. 真空，空白；
		v. 用真空吸尘器清扫
pulse	[pʌls]	n. 脉搏，脉冲；
		v. 搏动，跳动，震动
blast	[blɑːst]	n. 爆炸，（爆炸引起的）气浪，冲击波，突如其来的强劲气流；
		v.（用炸药）炸毁，爆破，严厉批评，猛烈抨击
hopper	['hɒpə(r)]	n. 形送料斗，漏斗
flue gas desulphurization		烟气脱硫
selective catalytic reduction		选择性催化还原
oxidation/reduction		氧化/还原
incinerator	[ɪnˈsɪnəreɪtə(r)]	n.（垃圾）焚化炉

whirl	[wɜːl]	v.（使）旋转，回旋，打转； n. 旋转，回旋，急转
centrifugal force		离心力
exponential	[ˌekspəˈnenʃl]	adj. 指数的，幂的，由指数表示的
diameter	[daɪˈæmɪtə(r)]	n. 直径，对径，放大率，放大倍数
pressure drop		压力降
cotton grin		轧棉机
rock crusher		碎石机，岩石压碎机
corrosion	[kəˈrəʊʒn]	n. 腐蚀，侵蚀
abrasive	[əˈbreɪsɪv]	adj. 有研磨作用的，研磨的； n.（用来擦洗表面或使表面光滑的）磨料
inlet	[ˈɪnlet]	n.（海、湖伸向陆地或岛屿间的）小湾，水湾，（液体、空气或气体进入机器的）入口、进口； v. 引进，嵌入，插入
plug	[plʌg]	n. 插头，（电源）插座，转换插头； v. 堵塞，封堵，补足，供给，推广
velocity	[vəˈlɒsəti]	n.（沿某一方向的）速度，高速，快速
transformer	[trænsˈfɔːmə(r)]	n. 变压器
voltage drop		电压降
charging electrode		充电电极
dislodge	[dɪsˈlɒdʒ]	v.（把某物）强行去除、取出、移动，（把某人）逐出、赶出、驱逐出
configure	[kənˈfɪgə(r)]	v.（按特定方式）安置，（尤指对计算机设备进行）配置，对（设备或软件进行）设定
propane	[ˈprəʊpeɪn]	n. 丙烷
microbial	[maɪˈkrəʊbɪəl]	adj. 微生物的，细菌的
metabolize	[məˈtæbəlaɪz]	v. 新陈代谢（将食物、矿物质等通过化学过程转换成新细胞、能量和废料）
spray chamber		喷淋室
lint	[lɪnt]	n.（织物在制作过程中从表面掉落的）纤维屑，飞花，（毛料、棉布等的）绒毛
tray tower		板式塔、托盘塔
sieve	[sɪv]	n. 滤器，筛子，笊篱，漏勺； v. 筛，过筛，滤

Notes

① After enough dust has built up on the filters, as indicated by a build up in pressure across the fabric, dust is periodically removed by blowing air back through the fabric, pulsing the fabric with a blast of air, or shaking the fabric.

参考译文：当袋式除尘器上积聚了足够的灰尘（如织物上积聚的压力所示）后，通过将空气回吹织物、用气流推动织物或摇晃织物来定期清除灰尘。

② The electrical field produced in the gas stream as it passes through the high voltage discharge introduces a charge on the particles, which is then attracted to the collecting plates.

参考译文：气流通过高压放电产生的电场时，电场会使颗粒物带电，然后，带电的颗粒物就被吸引到收集板上。

Exercises 1

1. According to the reading material, chose the best answer(s) from the options.

(1) Which of the followings can be used to control stationary gaseous pollutants?_____

A. Fabric filters   B. Biofilters   C. ESP   D. Cyclones

(2) The principal of cyclone is _____.

A. the centrifugal forces that cause the particles to be thrown against the walls of the cylinder and drop into a hopper below.

B. to destroy VOCs and odors by microbial oxidation of these problem compounds.

C. high-voltage discharged particles that are then attracted to the collecting plates.

D. thermal destruction functions

(3) _____ is added to convert NO_x, to nitrogen and water.

A. Oxygen   B. Nitrogen   C. Ammonia   D. Vapor

(4) In scrubbing, _____ is the most commonly used absorbent liquid.

A. ammonia   B. water   C. pollutant   D. particulate

(5) Which of the followings is most efficient to the removal of VOCs?_____

A. SCR   B. ESP   C. Biofilter   D. Absorption

2. Describe the principals of each control device in English, and link each device and targeted pollutants.

Part 2 Translation

英语翻译的词义选择

词是最小的、能够单独运用的语言单位，表示概念或概念间的相互关系。英语和汉语有着大致相同的词类，实词中都有名词、动词、形容词、副词、代词、数词，虚词中都有介词和连词。两种语言中都有象声词，所不同的是英语中有冠词，而汉语中有量词和语气词。不同的词类在句中充当不同的句子成分或起着不同的语法作用。

英汉两种语言都存在一词多类、一词多义、一词多用的现象，这给译者在实际操作中带来了巨大的困难。所谓一词多类就是指有的词可能属于好几种词类或词性，而在每个词性中的意思都可能会有所差异。所谓一词多义就是指有的词对应多个目的语中的词，译者也需要根据具体情况选择合适的词义来翻译。一词多用就是指一个词的用法应视其在句子中的具体情况（包括词类和词义两个方面）而定。凡此种种，无不需要译者认真加以考虑斟酌。如"热爱学习"和"学习外语"中，前者"学习"为名词，后者"学习"则是动词。

一、根据词的单复数确定词义

和汉语不同，英语的名词具有单数和复数之分。英语中有些名词的单复数同形，但表达的意思却不同，有些名词的单复数不同，但表达的意思则可能相同，也可能不同。所以在翻译过程中，根据词的单复数之分选择恰当的词义具有重要意义。例如：

(1) Electronics is a branch of science that deals with the study and application of electron devices.

译文：电子学是有关电子装置的研究和应用的一门科学。（electronics 看似为复数，实则单数，在句中为"电子学"的意思）

(2) Arthritis is a disease causing pain and swelling in the joints of the body.

译文：关节炎是引起人体关节疼痛和肿胀的疾病。（arthritis 是疾病的名称，不是复数。Joints 做复数时意为"关节"）

二、根据词的词类确定词义

上面已经说到了有些词作为不同的词类所具有的词义也有所不同，所以必须先确定其在句子中的词类，然后选择恰当的词义。例如：

(1) It is the atoms that make up iron, water, oxygen and the like.

译文：正是原子构成了铁、水、氧等类物质。（like 在句中作为名词，表示"相同之物"）

(2) In the sunbeam passing through the window there are fine grains of dust shining like gold.

译文：在射入窗内的阳光里，细微的尘埃像金子一样在闪闪发亮。（like 在句中做介词，表示"像"）

(3) Like charges repel, unlike charges attract.

译文：同性电荷相斥，异性电荷相吸。（like 在句中做形容词，表示"相同的"）

诸如此类的词很多，有介词做名词、动词之用，动词做名词、形容词等，在"词性的转换"一节将更加详细讨论此问题。

三、根据词的专业领域确定词义

现在各学科、各专业相互渗透，相互影响已是司空见惯了，这就使得词的使用范围更加广阔，一词多义现象也更加普遍。一词多义现象不仅包括同一个词在不同领域具有不同的意思，而且同一个词在同一个专业领域中意义也不同。

例如"fighter"一词，在"fighters of the PLA"中应作"中国人民解放军战士"理解，而在"fighter escort wing"中则作"护航战斗机联队"理解。又如"system"一词，在不同专业领域中也有不同的词义：在土木工程 pilling system（打桩设备）中意为"设备"，在电路 wiring system（电路图）中意为"图表"，在光学 optical system（光具组）中意为"组合"等。例如：

(1) Assume that the input voltage from the power supply remains constant.

译文：假定由电源输入的电压保持不变。（power 在句中表示"电源"）

(2) Combat power is useless unless it can be brought to bear at the right point, and at the right time.

译文：战斗力如果不能适时适地地发挥作用，这样的战斗力就等于零。（power 在句中表示能力的"力"）

(3) The combining power of one element in the compound must equal the combining power of the other element.

译文：化合物中一种元素的化合价必须等于另一种元素的化合价。（power 在句中表示化学中的"价"）

四、根据词语境确定词义

词与词是相互联系的，应把每个词放在句、段落乃至篇章更大的语言环境中加以考虑。我国学贯中西的著名学者林语堂曾说："字义是活的，每每因在文中的用法而变化。"例如：

(1) People who drink and smoke heavily are likely to develop cancers of the oral and tongue at a younger age.

译文：大量饮酒并吸烟的人很可能在年轻时就患上口腔癌和舌癌。（develop 在句中表示"患病"）

(2) In developing the design, we must consider the feasibility of processing.

译文：在进行设计时，必须考虑加工的可能性。（develop 在句中表示"设计"）

(3) In 1943, the British developed a new type of radar set which used a much shorter wavelength.

译文：1943年，英国人发明了一种波长短得多的新型雷达。（develop 在句中表示"研发、研制、发明"）

由于词、句等处于语言的大语境中，译者在理解和翻译过程中必须全盘考虑。此外，翻译还是个理解的全过程，在这个过程中，始终在进行对原文分析综合、判断和推理等认知活动。也就是说，理解是翻译的基础，在翻译的过程中，理解的连贯性、逻辑性是进行正确翻译的充分保证。

五、根据词的汉语习惯搭配确定词义

英语里有些词或词组的词义如何确定，也要根据汉语搭配习惯来选择。英、汉两种语言隶属于不同的语系（前者属于印欧语系，后者属于汉藏语系），在句子结构和表达形式上都有各自的特点。翻译的时候应注意考虑汉语的语言结构特点，力求用规范的汉语表达原文旨意。如果拘泥于原文意思的束缚而不考虑汉语的表达，译文也会不堪一读，从而影响对原文的理解，达不到理想的翻译效果。对于专业英语翻译则更应该从专业角度考虑语言的习惯搭配。例如：

Volcanic rocks from inside the field were found to be non-magnetic.

译文1：热田内的火山岩以前被发现是非磁性的。

译文2：以前的调查表明，这个热田的火山岩为非磁性岩石。

很显然，译文1在表达上不太符合汉语的表达习惯，句式怪异。相比之下，译文2的表达不仅符合汉语的表达，而且 found 一词翻译为"表明"，比翻译为"发现"更好，更符合科技英语中内容和表达的精确性。

另外，专业翻译不同于一般翻译，关键在于所要处理的材料是与专业有关的。也就是说，翻译出来的译文也要具有相应的专业性，也就是说要有"科技味"。比如，如果将 the great pressure and heat（高温高压）译为"巨大的压力和热"，将 advanced geophysics（高等地球物理学）译成"高级地球物理学"等，就没有了"科技味"，试问读者作何感想？以下再来看看一些搭配，例如：

build a ship 造船
build a bridge 架桥
universal meter 万用表
universal motor 交直流两用电动机
universal dividing heading 多用分度表头
universal constant 通用常数
universal rules 一般法则

六、根据词的同义词和替代词确定词义

英语和汉语相比，在同一个句子、段落、篇章中较少用相同词语，更喜欢用其同义词以避免重复，或用助动词 to，或用 one 来代替同一个意思的词，而汉语则喜欢用重复来加强表

达的力度，以达到活跃文风的效果。例如：

(1) ...combining production of wartime products with peace-time products and integrating military with civilian purposes.

译文：……国防工业实行军民结合、平战结合。（combine 和 integrate 两个同义词在句中都表示"结合"）

(2) Some people want to suspend the bombing, but others prefer not to.

译文：有些人想要停止轰炸，但其他人则认为还是不停止为好。（suspend 和 not to 在句中都表示"停止"）

(3) These weapons are better than those ones.

译文：这些武器比那些武器精良。（weapons 和 ones 都表示"武器"）

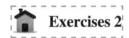

Exercises 2

Translate the following paragraph into Chinese.

(1) After enough dust has built up on the filters, as indicated by a build up in pressure across the fabric, dust is periodically removed by blowing air back through the fabric, pulsing the fabric with a blast of air, or shaking the fabric.

(2) Heavy dust at the inlet of the cyclone can also lead to plugging of the cyclone hopper.

(3) The process of adsorption involves the molecular attraction of gases or vapors [usually volatile organic compounds (VOCs)] onto the surface of certain solids (usually carbon, molecular sieves, and/or catalysts). This attraction may be chemical or physical in nature and is predominantly a surface effect. Activated carbon (charcoal), which possesses the large internal surface area needed to adsorb large quantities of gases within its structure, is often used to remove VOCs from flue gases.

Part 3　Writing

学术论文的撰写（二）——学术论文写作前的准备工作

英语科技论文写作对许多人来说是一项棘手的工作，但是只要把握好科技英语论文的三大要素和特点，把写作的过程分解成若干个步骤，写起来就容易多了。也就是说，把论文写作分解成较小的、可操作的若干个部分，再使用前面的章节中讲述过的技巧和方法，就可以使论文写作更加顺利。在开始论文写作前，可以从几个方面进行准备。

一、选择研究课题

选择研究课题是进行科学研究并撰写论文的基础。一般来说，准备研究课题时应注意以下三个方面的问题：

1. 听取他人的意见

听取他人的意见有利于在选择课题时做出正确的决定，特别是有更深学术背景和学术经验的导师的意见十分珍贵。这也说明了"听君一席话胜读十年书"。有时，与同辈或同行交谈也有可能激发灵感，从而产生一个非常有价值的研究课题。而且，在与他人的交谈中，可以不断获取新的想法，提出新的问题，讨论可行的解决问题的方法，让思考更加丰盈、更加深刻。

2. 根据个人的兴趣

兴趣是最好的老师。所以，在准备研究课题时，应该优先挑选那些自己感兴趣的或者选定之后有内在动力深入地研究下去的课题。

3. 头脑风暴

头脑风暴是为了在选题的过程中产生某种灵感或者想法。这实际上是一种自由联想的过程，也是一种对某一课题在排列其内容时有利于产生想法的方法。可以和同学、朋友等一起进行讨论，并将所有讨论出来的想法记录下来。

二、准备文献

准备文献实际上就是查阅、收集与课题有关的参考资料。一般来说，准备文献主要的来源有以下三个：

1. 查阅网络资源

现在利用互联网查找资料非常方便，可以很容易查找到与研究课题有关的资料。在线资料既有利于确定课题的题目，还有利于查找所需的信息。通过在网上搜寻相关的题目、论文和图书，对确定论文主题有极大的帮助。除了常用的搜索引擎外，很多学校和科研单位还会提供一些专业的学术搜索服务，具体可以在图书馆或文献中心获得相关方式。

2. 查阅图书资料

应该说查阅图书资料是确定研究课题最好的方法之一。图书通常对某个主题有完整、详细的研究方法和结果的介绍。有些图书还会提供与主题相关的详尽的背景资料，对充分地了解主题有很大的帮助。查阅图书的有效方法是，略读书名、前言、目录、索引，以确定是否有进一步仔细查阅的需要。

3. 查阅学术期刊

学术期刊通常是按月刊、双月刊或季刊的形式发行的，所含的信息、所提供的观点和报告都比较新颖。此外，由期刊提供的信息更加专业且内容也更详尽，这就更有利于进行相关的研究。使用期刊的文献索引可以很方便地查找到所需要的文章。其中最权威的索引有：SCI、SSCI、EI 和 ISTP 等。在查阅索引与文献时，只需要简单地查阅一下文章的题目和关键词就可以决定对论文的取舍。下面简单介绍一下重要的学术索引。

（1）SCI

《科学引文索引》（Science Citation Index，SCI）是美国科学情报研究所（ISI）出版的一种世界著名的期刊文献检索工具，也是当前世界自然科学领域基础理论学科方面的重要期刊文摘索引数据库。

SCI 是目前国际上三大检索系统中最著名的一种，其中以生命科学、医学、化学、物理所占比例最大，收录范围是当年国际上的重要期刊，尤其是它的引文索引表现出独特的科学参考价值，能反映自然科学研究的学术水平，在学术界占有重要地位。它主要收录文献的作者、题目、源期刊、摘要、关键词，不仅可以从文献引证的角度评估文章的学术价值，还可以迅速方便地组建研究课题的参考文献网络。利用它可以检索自 1945 年以来重要的学术成果信息。SCI 还被学术界当作制订学科发展规划和进行学术排名的重要依据。目前，SCI 涵盖学科超过 100 个，主要涉及农业、生物及环境科学，工程技术及应用科学，医学与生命科学，物理及化学，行为科学等。自然科学数据库有 5000 多种期刊，SCI 每年收集论文数达 60 万~70 万条。

（2）SSCI

《社会科学引文索引》（Social Sciences Citation Index，SSCI），是 SCI 的姊妹篇，亦由美国科学信息研究所创建，是目前世界上可以用来对不同国家和地区的社会科学论文的数量进行统计分析的大型检索工具。据 ISI 网站 2021 年公布的数据显示，SSCI 全文收录 3553 种世界最重要的社会科学期刊，内容覆盖包括人类学、法律、经济、历史、地理、心理学等 55 个领域。收录文献类型包括：研究论文、书评、专题讨论、社论、人物自传、书信等。选择收录期刊为 1900 多种，收录数据从 1956 年至今，是社会科学领域重要的期刊文摘索引数据库。数据覆盖了历史学、政治学、法学、语言学、哲学、心理学、图书情报学、公共卫生等社会科学领域。

（3）ISTP

《科学技术会议录索引》（Index to Scientific &Technical Proceedings，ISTP），是美国科学情报研究所的网络数据库（Web of Science Proceedings）中两个数据库（ISTP 和 ISSHP）之一。ISTP 创刊于 1978 年，它由美国科学情报研究所编辑出版，主要收录国际上著名的自然科学及技术方面的科技会议文献，包括一般性会议、座谈会、研究会、讨论会、发表会等的会议文献，涉及学科基本与 SCI 相同。ISTP 所收录的数据包括生命科学、农业、环境科学、生物化学、分子生物学、生物技术、医学、工程、计算机科学、化学、物理学、工程技术和应用科学等学科,其中工程技术与应用科学类文献约占 35%。1990—2003 年间,ISTP 和 ISSHP 共收录了 60 000 个会议的近 300 万篇论文的信息。ISTP 收录论文的多少与科技人员参加的重要国际学术会议多少或提交、发表论文的多少有关。我国科技人员在国外举办的国际会议上发表的论文占被收录论文总数的 64.44%。

（4）ISSHP

《社会科学和人文会议录索引》（Index to Social Sciences & Humanities Proceedings，ISSHP），于1979年由ISI创办，其数据涵盖了社会科学、艺术与人文科学领域的会议文献。这些学科包括哲学、心理学、社会学、经济学、管理学、艺术、文学、历史学、公共卫生等领域。

（5）EI

《工程索引》（Engineering Index，EI），是全世界最早的工程文摘来源，也是目前世界著名的工程技术类综合性检索工具。EI创刊于1884年，最初由美国工程情报公司(Engineering Information Co.)编辑出版发行。它主要收录工程技术领域的科技期刊和会议论文，其所收录文献的范围几乎覆盖工程技术各个领域的数据，涉及材料工程、地质、电工、电子、通信、动力、核技术、化学、工业工程、环境、机械工程、计算机和数据处理、交通运输、金属工艺、控制工程、矿冶、能源、材料科学、农业、食品技术、汽车工程、生物工程、石油、食品、数理、水利、土木工程、医学、仪表、应用物理、宇航、照明、光学技术和自动控制等学科领域。EI数据库每年新增约50万条文摘索引信息，分别来自5100种工程期刊、会议文集和技术报告，其中大约22%为会议文献，90%文献的语种为英文。EI每月出版1期，每期文摘1.3万~1.4万条，年报道文献量16万余条。EI具有综合性强、资料来源广、地理覆盖面广、报道量大、报道质量高和权威性强等特点。1992年，EI公司开始收录中国期刊，并于1998年在清华大学图书馆建立了EI中国镜像站。

（6）CSSCI

《中文社会科学引文索引》（Chinese Social Science Citation Index，CSSCI），是南京大学1997年在全国率先提出并研制的一种中文社会科学研究信息检索的产品。CSSCI遵循文献计量学规律，采取定量与定性评价相结合的方法，从全国2800余种中文人文社会科学学术性期刊中精选出学术性强、编辑规范的期刊作为来源期刊。现已开发的CSSCI收录了1998年以来的数据，其来源文献54万余篇，引文文献320余万篇。该项目成果填补了我国社会科学引文索引的空白，达到了国内领先水平。目前，教育部已将CSSCI数据作为全国高校机构与基地评估、成果评奖、项目立项、名优期刊的评估、人才培养等方面的重要指标。

（7）CSCD

《中国科学引文数据库》（Chinese Science Citation Database，CSCD），属教育部主管，清华大学主办，是由中国学术期刊（光盘版）电子杂志社创办的我国学术期刊全文检索与评价数据库。CSCD收录我国数学、物理、化学、天文学、地学、生物学、农林科学、医药卫生、工程技术、环境科学和管理科学等领域出版的中英文科技核心期刊和优秀期刊近千种，其中核心库来源期刊670种，扩展库期刊为378种，已积累了从1989年至今的论文记录近100万条，引文记录近400万条。CSCD分为核心库和扩展库，共遴选了1048种期刊。

三、做好笔记

在大量查阅文献的同时要做好笔记，以便记录必要的信息和资料的出处。同时，在阅读相关的资料时，要根据对所读到的资料的理解和启发来组织写文章的思路。精心做好的笔记应包括作者原始的观点、基本的事实和可能在论文或著作中要使用的直接引语。毋庸置疑，做好笔记可以使论文写作更为容易和方便。

做笔记的方式各种各样，可以用纸笔做笔记，也可以用电脑或其他电子媒介。使用电脑或其他电子媒介便于储存和检索，也便于按需进行整理和使用。

1. 选择有价值的资料

在阅读资料和做笔记前，很有必要对所获得的资料进行评价，选择有价值的资料进行记录。可以从三个方面进行筛选：

（1）作者的姓名、著作的书名、出版社、出版日期等。因为这样既可表明是否与要做的课题、要撰写的论文有关，还可以发现这位作者是否是某一领域的权威，书中所含信息是否是最新的。一般来说，如果其他所有的要素都是一样的，那么，在科技领域首选的资料在日期上应是最近的。

（2）要注意资料的来源，即所获取的资料是原始资料还是二手资料，作者是在提供其个人的经历和发现还是在重复他人的发现和经历。应该说，原始资料要比二手资料可靠得多。

（3）要注意资料来源的客观性。资料的来源是带有偏见还是非常客观对于研究和撰写论文非常重要。通常，学术期刊和大学学报以及一些独立机构的出版物都是比较客观的。

2. 笔记的内容

笔记的内容通常包括以下五个方面：

（1）背景资料；

（2）支持预定课题观点的总结；

（3）解释性信息，如定义、内容层次上的小结以及按年代排列的数据；

（4）直接引语、实例、事实与掌故；

（5）统计数据，如比例、质量、数量、比值与日期等。

3. 做笔记的方法

（1）要记录资料的来源。在记笔记之前，首先要将作者的姓名、文章的题目、出版地、出版社、出版日期以及版权日期、选中的内容的页码都记下来。

（2）全面且简要。在记笔记时既不要写得太多也不要写得太少，只需把那些与研究课题有关的而又在其他渠道无法获取的内容记下来。记笔记时一定要忠实于原作的观点。

（3）一页一项。用纸笔做笔记时，在一张卡片或一页上只记录一项相关的内容，这样有利于整理内容。

（4）在电脑上做笔记时，也要像在纸上做笔记一样，尽量"一页一项"，并在每个文件上标注一些关键词，并把有相同关键词的笔记文件归类到一个文件夹里，这样有助于检索。

 **Exercises 3**

Searching and reading literatures (no less than 10 papers) about the recent development of the removal of VOCs in environmental engineering field by SCI, EI and ISTP.

 Expanding Reading

Rising Emissions Overshadow Airlines' Fuel-Efficiency Gains

New mandates from aviation authorities will not go far enough to reduce greenhouse gases, experts say

By Maxine Joselow

Despite modest gains in fuel efficiency, domestic airlines continue to pump larger and larger quantities of planet-warming pollution into the atmosphere.

Airlines increased their fuel efficiency by 3 percent on average last year, according to a new report from the International Council on Clean Transportation. But overall, the gains were not enough to offset rising greenhouse gas emissions from the domestic industry, which released 7 percent more carbon dioxide into the atmosphere than last year.

"A big takeaway is that despite voluntary efforts from the airlines to improve their fuel efficiency, CO_2 emissions from air travel continue to grow rapidly," said Dan Rutherford, program director for marine and aviation at the ICCT and a co-author of the report.

The report looked at 13 airlines that operated in the United States in 2018, including Alaska Airlines, American Airlines Inc., Delta Air Lines Inc., Frontier Airlines and United Airlines Inc.

Frontier performed the best, improving its fuel efficiency by 4 percent through investments in newer planes and more direct routes.

But other airlines were laggards. JetBlue Airways Corp. performed at the bottom of the pack, burning 26 percent more fuel than Frontier on comparable flights last year.

A proliferation of new flights—and seats available on those flights—has been a major contributor to rising carbon dioxide emissions.

As airlines continued to add new routes last year, passengers traveled 10 percent more miles and airlines burned 7 percent more fuel, according to the report.

"In 2014, there was that big oil price collapse, and fuel got cheaper," Rutherford said. "The U.S. airlines reacted by increasing the number of flights and seats available. And customers really responded to that."

GOALS AND RULES

The report comes as the Federal Aviation Administration works toward a goal of carbon-neutral growth for the industry by 2020. The goal dates back to 2015, when President Obama was in office. But it remains in effect under President Trump, whose administration has sought to roll back or delay Obama-era environmental regulations for coal plants and cars.

"Probably this administration would like to forget about it," Rutherford said. "But right now, it's still on the books."

Perhaps surprisingly, EPA is preparing to issue the first-ever greenhouse gas emissions limits for planes (Climatewire, May 29).

The agency is mulling the new carbon rules in response to a forthcoming mandate from the International Civil Aviation Organization. ICAO, a specialized body of the United Nations, proposed international limits on aircraft carbon emissions in 2017. The limits are slated to take effect for new planes on Jan. 1, 2020, and for existing planes in 2028.

Still, clean aviation experts view the ICAO rules as too lenient.

"The ICAO standards themselves are pretty weak," Clare Lakewood, a senior attorney at the Center for Biological Diversity, previously told E&E News.

"They're basically an anti-backsliding provision," Lakewood said. "They require CO2 reductions of 4 percent over 12 years. The market forces alone are predicted to achieve efficiency gains of about 10 percent."

Rutherford agreed.

"From our perspective, we need not only a new standard, but a more stringent standard," he said.

Reprinted from Climatewire with permission from E&E News on September 12, 2019. E&E provides daily coverage of essential energy and environmental news at www.eenews.net.

(https://www.scientificamerican.com/article/rising-emissions-overshadow-airlines-fuel-efficiency-gains/)

Unit 3 Water Pollution and Control

Lesson 1 Processing of Water Supply

Part 1 Reading

The objective of municipal water treatment is to provide a potable supply— that is chemically and microbiologically safe for human consumption. For domestic uses, treated water must be aesthetically acceptable—free from apparent turbidity, color, odor, and objectionable taste. Quality requirements for industrial uses are frequently more stringent than for domestic supplies. Thus, additional treatment may be required by the industry. For example, boiler feed water must be demineralized to prevent scale deposits. Common water sources for municipal supplies are groundwater, rivers, natural lakes, and reservoirs.

Pollution and eutrophication are major concerns in surface-water supplies. Water quality depends on agricultural practices in the watershed, location of municipal and industrial outfall sewers, river development such as dams, season of the year, and climatic conditions. Periods of high rainfall flush silt and organic matter from cultivated fields and forest land, while drought flows may result in higher concentrations of wastewater pollutants from sewer discharges[①]. River temperature may vary significantly between summer and winter. The quality of water in a lake or reservoir depends considerably on season of the year. Municipal water quality control actually starts with management of the river basin to protect the source of water supply. Highly polluted waters are both difficult and costly to treat. Although some communities are able to locate groundwater supplies, or alternate less polluted surface sources within feasible pumping distance, the majority of the nation's surface supplies are facing the thread of contamination. The challenge in waterworks operation is to process these waters to a safe product acceptable for domestic use.

The primary process in surface-water treatment is chemical clarification by coagulation, sedimentation, and filtration, as illustrated in Figure 1. Lake and reservoir water have a more uniform year-round quality and require a lesser degree of treatment than river water. Natural purification results in reduction of turbidity, coliform bacteria, color, and elimination of day-to-day variations. On the other hand, growths of algae cause increased turbidity and may produce difficult-to-remove tastes and odors during the summer and fall. The specific chemicals applied in coagulation for turbidity removal depend on the character of the water and economic considerations. The most popular coagulant is alum (aluminum sulfate). As a flocculation aid, the common auxiliary chemical is a synthetic polymer. Activated carbon is applied to remove taste and odor producing compounds. Chlorine and fluoride are post-treatment chemicals. Prechlorination may be used for disinfection of the raw water only if it does not result in formation of trihalomethanes (THMs)[2].

River supplies normally require the most extensive treatment facilities with greatest operational flexibility to handle the day-to-day variations in raw water quality. The preliminary step is often presedimentation to reduce silt and settleable organic matter prior to chemical treatment. As illustrated in Figure 3.1, many river-water treatment plants have two stages of chemical coagulation and sedimentation. As many as a dozen different chemicals may be used under varying operating conditions of provide a satisfactory finished water.

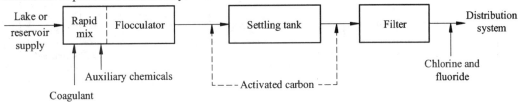

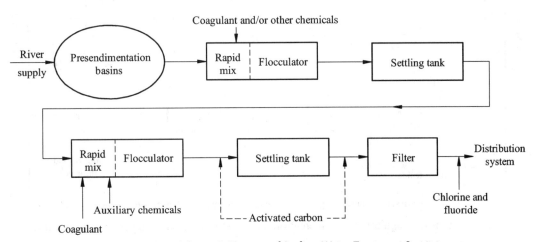

Figure 3.1 Schematic Patterns of Surface-Water Treatment Systems

Well supplies normally yield cool, uncontaminated water of uniform quality that is easily processed for municipal use. Processing may be required to remove dissolved gases and undesirable minerals. The simplest treatment illustrated in Figure 3.2(a) is disinfection and fluoridation. Deep-well supplies may be chlorinated to provide residual protection against potential contamination in the water distribution system. In case of shallow wells recharged by surface waters, chlorination can both disinfect the groundwater and provide residual protection. Fluoride is added to reduce the incidence of dental caries. Dissolved iron and manganese in well water oxidize when contacted with air, forming tiny rust particles that discolor the water. Removal is performed by oxidizing the iron and manganese with chlorine or potassium permanganate, and removing the precipitates by filtration [Figure 3.2(b)].

Excessive hardness is commonly removed by precipitation softening, shown schematically in Figure 3.2(c). Lime and, if necessary, soda ash are mixed with well water, and settleable precipitate is removed. Carbon dioxide is applied to stabilize the water prior to final filtration. Aeration is a common first step in treatment of most groundwater to strip out dissolved gases and add oxygen.

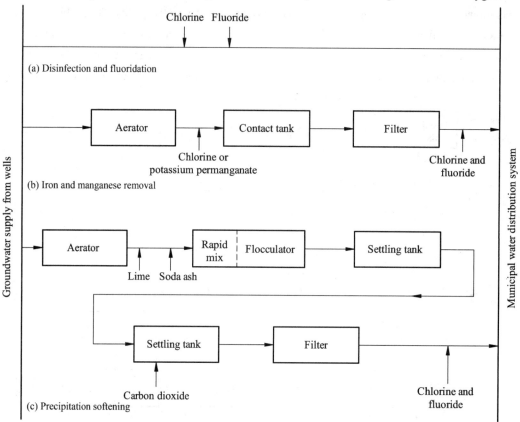

Figure 3.2   Flow Diagrams of Typical Groundwater Treatment Systems

Clarification by coagulation, sedimentation, and filtration removes suspended impurities and turbidity from drinking water. The final step is disinfection which produces potable water, free of harmful microorganism. But other treatment processes may be required, particularly to remove some of the dissolved substances. These processes may be used in addition to clarification or applied separately, depending on the source and quality of the raw water

Groundwater, for example, does not ordinarily require clarification because the water is filtered naturally in the layers of soil from which it is withdrawn. Disinfection of groundwater supplies, required by law for public water supply systems, is basically a precautionary step;

groundwater is usually free of bacteria or other microorganisms. On the other hand, because of its contact with soil and rock, groundwater may contain high levels of dissolved minerals that must be removed.

The two primary sources of waste from water treatment processes are sludge from the settling tank, resulting from chemical coagulation or softening reactions, and wash water from backwashing filters. These discharges are highly variable in composition, containing concentrated materials removed from the raw water and chemicals added in the treatment process. The wastes are produced continuously, but are discharged intermittently. Historically, the method of waste disposal was to discharge to a watercourse or lake without treatment. This practice was justified from the viewpoint that filter backwash waters and settled solids returned to the watercourse added no new impurities but merely returned material that had originally been present in the water. This argument is no longer considered valid, since water quality is degraded to the extent that a portion of the water is withdrawn, and chemicals used in processing introduce new pollutants. Therefore, more stringent federal and state pollution control regulations have been enacted requiring treatment of waste discharges from water purification and softening facilities.

The situation of each waterworks is unique and controls to some extent the method of ultimate disposal of plant wastes. For example, they may be piped to a municipal sewer for processing or may be discharged to lagoons, provided that sufficient land area is available. Ultimate disposal by landfill, or barging to sea, requires thickening for economical handling and hauling. A variety of alternative processing methods are available; however, because of unique characteristics of each plant's waste, no specific process can be universally applied.

Words and Expressions

domestic	[dəˈmestɪk]	adj. 民用的，本国的，国内的，家务的
aesthetically	[iːsˈθetɪkəlɪ]	adv. 审美地，美学观点上地
turbidity	[tɜːˈbɪdɪtɪ]	n. 浊度，浑浊度
stringent	[ˈstrɪndʒənt]	adj. 严格的，严厉的，紧缩的，短缺的
boiler	[ˈbɔɪlə(r)]	n. 锅炉，汽锅
deposit	[dɪˈpɒzɪt]	n. 订金，订钱，押金，存款；v. 使沉积，使沉淀，使淤积，放下，放置，将（钱）存入银行，存储
municipal supply		市政供水
demineralize	[ˌdiːˈmɪnərəlaɪz]	v. 脱盐，除去（水）的盐
scale	[skeɪl]	n. 水垢，（尤指与其他事物相比较时的）规模，比例尺，等级，鱼鳞，音阶，称，天平；

			v. 攀登，翻越，去鳞，刮除牙石
eutrophication	[ˌjuːtrəfɪˈkeɪʃn]		n. （水体的）富营养化
surface-water			n. 地表水
watershed	[ˈwɔːtəʃed]		n. 流域，转折点，分界线，分水岭（标志着重大变化的事件或时期）
flush	[flʌʃ]		v. （用水）冲洗净，冲洗，脸红，冲（抽水马桶）；
			n. 脸红，潮红，一阵强烈情感，（流露出的）一阵激情，冲（抽水马桶）；
			adj. 富有，很有钱（通常为短期的）
drought	[draʊt]		n. 干旱，旱灾
pumping distance			泵送距离
waterworks	[ˈwɔːtəwɜːks]		n. 自来水厂，（人体的）泌尿系统
clarification	[ˌklærəfɪˈkeɪʃn]		n. 澄清（法），净（纯）化，清化（理），说（阐）明，解释
coagulation	[kəʊˌæɡjuˈleɪʃn]		n. 絮（胶）凝，凝结（物），凝聚，凝固（作用），聚集
sedimentation	[ˌsedɪmenˈteɪʃn]		n. 沉积（作用）
filtration	[fɪlˈtreɪʃn]		n. 过滤，滤清，滤除
purification	[ˌpjʊərɪfɪˈkeɪʃ(ə)n]		n. 提纯，清洗，净化（作用），斋戒
coliform	[ˈkoʊləˌfɔrm]		n. 大肠（埃希氏）杆菌
elimination	[ɪˌlɪmɪˈneɪʃn]		n. 去除，消除，消灭，淘汰，淘汰赛
aluminum sulfate			n. 硫酸铝
alum	[ˈæləm]		n. 明矾，白矾（用于制革、印染等）
algea	[ˈælˌdrɪə]		n. 藻类植物
flocculation aid			n. 助凝剂
auxiliary	[ɔːɡˈzɪliəri]		n. 助动词，辅助，辅助人员；
			adj. 辅助的，备用的
synthetic polymer			n. 合成高分子材料，合成聚合物
chlorine	[ˈklɔːriːn]		n. 氯，氯气
fluoride	[ˈflɔːraɪd]		n. 氟化物
raw water			n. 原水
trihalomethanes	[trɪhæləʊmeˈθeɪnz]		n. 三氯甲烷
flexibility	[ˌfleksəˈbɪləti]		n. 灵活性，弹性，柔韧性，适应性
preliminary	[prɪˈlɪmɪnəri]		adj. 初步的，预备性的，开始的；
			n. 初步行动（或活动），预备性措施
silt	[sɪlt]		n. （沉积的）泥沙，淤泥，粉沙

finished water		n. 出厂水
uncontaminated	[ˌʌnkən'tæmɪneɪtɪd]	adj. 未被损害的，未受污染的
mineral	['mɪnərəl]	n. 矿物质，矿物，汽水
disinfection	[ˌdɪsɪn'fɛkʃən]	n. 消毒
dental	['dentl]	adj. 牙齿的，牙科的，齿音的
caries	['keərɪːz]	n. 龋齿，骨疡，骨疽
manganese	['mæŋɡəniːz]	n. 锰
precipitate	[prɪ'sɪpɪteɪt]	n. 沉淀物，析出物；v. 使……突然降临，加速（坏事的发生），使突然陷入（某种状态）
rust	[rʌst]	n. 锈，铁锈，（植物的）锈病，锈菌；v. （使）生锈
potassium permanganate		n. 高锰酸钾
hardness	['hɑːdnɪs]	n. 硬（刚，强）度，（坚）硬性，坚固（牢），核防护能力，困难，苛刻
schematically	[skɪː'mætɪkəlɪ]	adv. 示意性的，按照图式，计划性地
lime	[laɪm]	n. 石灰，酸橙
soda-ash	['səʊdə æʃ]	n. 纯碱，苏打粉，碳酸钠
aeration	[eə'reɪʃn]	n. 曝气，通（鼓）风
dissolved substance		n. 可溶物，溶解物
intermittently	[ˌɪntə'mɪtəntlɪ]	adv. 间歇地
degrade	[dɪ'ɡreɪd]	v. （使）降解，分解，降低……身份，使受屈辱，降低，削弱（尤指质量）
lagoon	[lə'ɡuːn]	n. （人造）污水储留池，潟湖，环礁湖
landfill	['lændfɪl]	n. 废物填埋场，填埋的废物
recarbonation	['riːkɑːbə'neɪʃn]	n. 再碳酸化
discharge	[dɪs'tʃɑːdʒ]	v. 释放，排出，放出，流出，准许（某人）离开，解雇；n. 排出（物），放出（物），流出（物），获准离开，免职，出院，退伍
backwash	['bækwɒʃ]	n. 反冲洗，（行船激起的）尾流，（波浪拍击岸边后的）退浪，恶果，余波
softening facilities		n. 软化设备
ultimate disposal		n. 最终处置
haul	[hɔːl]	v. （用力）拖、拉、拽，用力缓慢挪动到（某处），强迫（某人）去某处

Notes

① Periods of high rainfall flush silt and organic matter from cultivated fields and forest land, while drought flows may result in higher concentrations of wastewater pollutants from sewer discharges.

参考译文：丰水期大量降雨会将农田和林地的淤泥和有机物冲刷到地表水域，而在枯水期排水系统排放的废水则可能导致地表水域污染物浓度更高。

② Prechlorination may be used for disinfection of the raw water only if it does not result in formation of trihalomethanes (THMs).

参考译文：只有在未形成三氯甲烷（THMS）的情况下，预氯化才能用于原水消毒。

注：饮水中的三氯甲烷大部分是经过有机物氯化而形成的。根据 EPA 调查发现，加氯处理后的饮用水 95%～100%含有三氯甲烷，平均浓度为 20 μg/L。三氯甲烷属中等毒性，主要作用于中枢神经系统，具有麻醉作用，对心、肝、肾有损害。

Exercises 1

1. According to the reading material, chose the best answer(s) from the options.

(1) The objective of municipal water treatment is to _____.

A. provide a safe supply for environment.

B. make all waste water microbiologically safe.

C. provide a potable supply for human consumption.

D. meet the domestic use.

(2) Surface water quality depends on and industrial outfall sewers, river development such as dams, _____, _____, _____ and _____.

A. agricultural practices in the watershed B. location of municipal

C. season of the year D. climatic conditions

(3) The primary process in surface-water treatment is chemical clarification by _____.

A. coagulation B. disinfection C. sedimentation D. filtration

(4) The preliminary step is often _____ to reduce silt and settleable organic matter prior to chemical treatment.

A. clarification B. presedimentation C. disinfection D. filtration

(5) Which statement is FALSE? _____

A. The final step of process of river supplies is disinfection

B. Groundwater, does not ordinarily require clarification and disinfection while it may contain high levels of dissolved minerals that must be removed.

C. Clarification by coagulation, sedimentation, and filtration removes suspended impurities and turbidity from drinking water.

D. The two primary sources of waste from water treatment processes are sludge and the settling tank.

2. Propose how to process the surface water supplies.

3. Propose how to process the ground water supplies.

Part 2　Translation

英语翻译的词义引申

在科技英语的翻译中，不仅会对词义进行变化，有时还会对词义进行引申，因为在翻译过程中，有一些源语言词汇难以匹配目的语，所以，结合上下文语境，对词义进行引申，能够更好地帮助读者理解原文想要表达的意思。一般来说，译者只有在当原有词义不能准确表达原意，或采用原有词义会造成读者误读或曲解时才选择这种处理方法。引申的有单个的词，也有词组或短语，必须视具体情况而定。以下就从词义具体化和词义抽象化两个方面来加以讨论。

一、词义具体化引申

有时，为了使词义所表达的含义更加精确严密，针对源语言中意思较为概括、笼统和抽象的词，根据目的语表达的需要，应对其含义做一番必要的调整、变动和延伸，使其具体化并易于为读者接受。一般来说，做词义具体化的引申有两种情况：一是将英语中代表抽象概念或属性的词用以具体化引申表示一种具体事物；二是有些词在特定的上下文中，其意义清晰，但译成汉语就必须具体化引申，否则难以理解接受，例如：

(1) By experimentation, imagination, and reasoning, mathematics are discovering new facts and ideas that science and engineering are using to change our civilization.

译文：数学家们通过实验、想象和推理，正在不断发现各种新事实和新观念，以便科学家和工程师可以用来改造我们的文明世界。

"mathematics"一词为"数学"，将其具体化引申为"数学家"；"science and engineering"具体化引申为"科学家和工程师"；"civilization"一词为"文明"，将其具体化引申为"文明世界"。

(2) Periods of high rainfall flush silt and organic matter from cultivated fields and forest land, while drought flows may result in higher concentrations of wastewater pollutants from sewer discharges.

译文：丰水期大量降雨会将农田和林地的淤泥和有机物冲刷到地表水域，而在枯水期排水系统排放的废水则可能导致地表水域污染物浓度更高。

"Periods"一词为"一段时期",根据上下文得知这段是用来描述地表水水质的季节变化,因此,结合下文的"drought flows",引申为"丰水期"。

二、词义抽象化引申

词义抽象化是和词义具体化相对应的。在英语中,人们常常会用一个表示具体形象的词来表示一种属性或一类事物。因此,在翻译时我们有时也将其词义进行必要的抽象化和高度概括性的引申。例如:

(1) The contributors in component technology have been in the semi-conductor component.

译文:元件技术中起主要作用的因素是半导体元件。

"contributor"一词为"捐款人,捐助者,投稿者",而在此句中表示起作用的"因素"。

(2) Another factor in the development of dermatology has been the growth of societies, regional and national, at which much cross-fertilization of minds took place.

译文:皮肤病学获得进展的另一个原因就是地区性和全国性皮肤病学会的成立。在这些学会中,人们得以彼此自由地交流想法。

"cross-fertilization"一词为"异花受精"或"以有益的方面影响某人或某事",其英文定义为"stimulate sb./sth. usefully or positively with ideas from a different field",所以根据"cross-fertilization of minds"的词义限定,可以处理为"交流想法、交流思想"。

(3) A small group of men were standing near a strange flight machine, one of them whose name was Wilbur Wright, began to push the machine along a rail.

译文:一小群人站在一架奇特的飞行器旁边。其中一个名叫威尔伯·赖特的人开始沿着铁轨推动飞行器。

"machine"一词为"机器",根据句中提到的"flight machine",我们知道怀特是在试验发明飞机,所以联想起来我们可以将其处理为"飞行器"。

(4) Scientists make experiments in order to look into, under, and behind things for answers they are asking.

译文:科学家们进行实验,对事物做探讨、研究、验证,以寻求对他们所提问题的答案。

"look into"为"看进去,朝……里面看去","look under"为"朝……下面看","look behind"为"向……背后看",在句中,译者将其分别抽象化为"探讨""研究""验证"。

除了单个词的引申外,科技英语中还有很多词组、习惯表达和搭配的引申。例如:

(5) Malaysia, which posted its highest growth rate in a decade, is the region's new star performer, with Indonesia close on its heels.

译文:10年中,马来西亚的发展速度最快。它是该地区新出现的最出色的经济发展国,印度尼西亚紧随其后。

"star"用来指"因出演成功而名扬四方的演员或歌星","star performer"当然也可指"扮演最重要角色的演员"或"领衔主角"。通过整个句子,我们知道,"star performer"与东南亚地区各国的经济发展有关,可将其引申为"最出色的经济发展国",符合语境。

(6) Hungary joined the growing list of East European nations turning their backs on Soviet aircraft-makers.

译文：东欧越来越多的国家不再订购苏联制造的飞机，匈牙利也是其中之一。

"turn one's back on something or someone"意为"背弃""抛弃""轻视"或"拒绝"某种事物或某人，根据原文语境，可将其引申为"不再订购（苏联制造的飞机）"，符合句意。

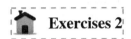 **Exercises 2**

Translate the following paragraph into Chinese.

Softening by ion exchange can produce water with almost zero levels of hardness, but this is not really desirable. Very soft water may be aggressive, or corrosive, causing damage to metal pipes and plumbing. Hardness levels of about 100 mg/L are considered optimum for drinking water. There is also some evidence that the presence of moderate hardness levels in drinking water actual reduces the incidence of heart disease. Another factor that must be considered is that softened water from an ion exchanger contains sodium, which may be harmful to who already have heart disease. In such cases, the softened water may not be suitable for consumption. Finally, it should be noted that ion-exchange softening does not produce a precipitate or sludge and is generally less costly than lime-soda softening. But because of the disadvantages mentioned, it is usually better adapted for treating industrial water supplies or for use in individual home softening units.

Part 3　Writing

学术论文的撰写（二）——学术论文写作前的准备工作

在撰写学术论文前的准备工作中，学会阅读学术论文是至关重要的。这里选用施一公教授发表的一篇文章，为学习学术论文阅读的学生指出努力的方向。

学生如何提高专业英文阅读能力

施一公

【序：此文针对本科生和一、二年级的博士、硕士生，对高年级的博士生和博士后也应该有参考价值。】

从小到大，我感性思维多一些，不善于读书。1985 至 1989 年在清华生物系读本科期间，从未读过任何一种英文专业期刊。我受到的与英文阅读相关的训练一共只有两个。一是我在 1986 年暑假期间选修的时任系主任的蒲慕明老师开设的《生物英语》系列讲座，隐隐约约记得蒲先生让我们阅读一些诸如 DNA 双螺旋发现之类的科普性英文文章，很有意思。但时间较短，暑假过后也没有养成读英文文章的习惯。二是《生物化学》这门课。与现在的清华生命学院形成鲜明对比，我上大学期间的所有基础课和专业课都是采用中文教材、中文讲课，只有郑昌学老师讲授的《生物化学》采用了 Lehninger 的 *Principles of Biochemistry*，而且郑老师要求我们每个学生每次课后阅读 10~20 页教材。我们同学大多感觉到专业英文阅读有所提高。

1990 年 4 月至 7 月初，我在依阿华州 Ames 小镇的 Iowa State University 度过了初到美国的前三个月，其中大部分时间在 Herbert Fromm 教授的实验室做轮转（rotation），跟随刘峰和董群夫妻两人做研究（刘峰现在 University of Texas Health Science Center 做教授）。当时感觉最困难的就是读专业论文。有一次，Fromm 教授要求我在组会上讲解一篇 *Journal of Biological Chemistry* 的文章，我提前两天开始阅读，第一遍花了足足六个小时，许多生词只能依靠英汉词典，文章中的有些关键内容还没有完全读懂，当时的感觉是 JBC 的文章怎么这么长、这么难懂？真有点苦不堪言。为了能给 Fromm 教授和师兄师姐留下好印象，第二天又花了好几个小时读第二遍，还做了总结。第三天我在组会上的表现总算没有给清华丢脸。但是，前前后后，真搞不清楚自己为了这一篇文章到底花了多少时间！

1990 年 7 月我转学到约翰霍普金斯大学以后，与本科来自北大的虞一华同在 IPMB program。虞一华大我一岁，来巴尔地摩之前已经在夏威夷大学读了一年的研究生，对于科研论文的阅读比我强多了。他常常在 IPMB 的办公室里拿着《科学》和《自然》周刊津津有味地阅读，看得我很眼馋，也不理解其中那些枯燥的文章有什么意思。他告诉我：他在读很有意思的科学新闻。科学新闻能有什么意思？虞一华给我讲了好几个故事：洛克菲勒大学校长、诺贝尔奖得主 David Baltimore 如何深陷泥潭，人类基因组测序如何争辩激烈，HIV 病毒究竟是谁发现的等。我还真没有想到学术期刊上会有这么多我也应该看得懂的内容！从那时起，每一期新的《科学》和《自然》一到，我也开始尝试着阅读里面的新闻和研究进展介绍，这些内容往往出现在"News & Comment" "Research News" "News & Views" "Perspectives"等栏目，文笔平实，相对于专业的科研论文很容易读懂。有时，我还把读到的科研新闻讲给我的同事朋友们听，而同事的提问和互动对我又是更好的鼓励。除了《科学》和《自然》，我也常常翻看《科学美国人》（*Scientific American*）。

与《细胞》（*Cell*）、《生物化学期刊》（JBC）等非常专业的期刊不同，《科学》和《自然》里面有相当一部分内容是用来做科普教育的。《科学》周刊的"Perspectives"和《自然》周刊的"News & Views"栏目都是对重要科学论文的深入浅出的介绍，一般 1~3 页，读起来比较通俗易懂，较易入门。读完这些文章后，再读原始的科学论文，感觉好多了！而且可以把自己的体会与专家的分析比较一下，找找差距，有时甚至也能找回来一点自信！

从 1998 年在普林斯顿大学任职到现在清华大学做教授，我总是告诉自己实验室的所有年轻人（包括本科生、硕士生、博士生、博士后）下面这几点读科研论文的体会，也希望我的学生跟我学：

1. 请每位学生每周关注《科学》和《自然》。（生命科学界的学生还应该留心《细胞》）。如果时间有限，每周花一个小时读读这两种周刊里的文章标题以及与自己研究领域相关的科研论文的 abstract 即可！这样做可以保证一个学生基本上能够跟踪本领域最重要的发现和进展，同时开阔视野，大概知道其他领域的动态。

2. 在时间充足的情况下，可以细读《科学》和《自然》里的新闻及科研论文。如果该科研论文有"News & Views"或"Perspectives"来介绍，请先读这些文章，这类导读的文章会提炼问题，就好比是老师事先给学生讲解一番论文的来龙去脉，对学生阅读原始论文有很大帮助。

3. 在读具体的科研论文时，最重要的是了解文章的主线逻辑。文章中的所有 Figures 都是按照这个主线逻辑展开描述的。所以，我一般先读"introduction"部分，然后很快地看一遍 Figures。大概知道这条主线之后，才一字一句地去读"results"和"discussion"。

4. 当遇到一些实验或结果分析很晦涩难懂时，不必花太多时间深究，而力求一气把文章读完。也许你的问题在后面的内容中自然就有解答。这与听学术讲座非常相似！你如果想每个细节都听懂，留心每一个技术细节，那你听学术讲座不仅会很累，而且也许会为了深究一个小技术环节而影响了对整个讲座逻辑推理及核心结论的理解。

5. 对个别重要的文章和自己领域内的科研论文，应该精读。对与自己课题相关的每一篇论文则必须字斟句酌地读。这些论文，不仅要完全读懂，理解每一个实验的细节、分析、结论，还必须联想到这些实验和结论对自己的课题的影响和启发，提出自己的观点。

6. 科学论文的阅读水平是循序渐进的。每个人开始都会很吃力，所以你有这种感觉不要气馁。坚持很重要，你一定会渐入佳境。当你有问题时或有绝妙分析时，应该与师兄师姐或找导师讨论。

7. 科研训练的一个重要组成部分就是科研论文的阅读。每一个博士生必须经过严格的科研论文阅读的训练。除了你自己的习惯性阅读外，你应该在研究生阶段选修以阅读分析专业文献为主的一至两门课，在实验室内也要有定期的科研论文讨论（Journal Club）。如果你的实验室还没有这种讨论，你们学生可以自发地组织起来。

8. 前面几条都是讨论如何提高科研论文的阅读能力，但是一旦入了门，就要学会 critical reading。不要迷信已发表的论文，哪怕是发表在非常好的期刊上。要时刻提醒自己：该论文逻辑是否严谨，数据是否可靠，实验证据是否支持结论，你是否能想出更好的实验，你是否可以在此论文的基础上提出新的重要问题？等等。

天外有天，读科研论文是一件很简单、但也很深奥的事情。一般的学生常常满足于读懂、读透一篇好的论文，优秀的学生则会举一反三，通过查找 references 纵深了解整个领域的历史、现状，并展望该领域未来的可能进展。

我从 1990 年对学术论文一窍不通到 1996 年博士后期间的得心应手，还常常帮助同事分析，自以为水平了得。但是有一件事让我看到了自己的严重不足，颇为羞愧。1996 年，是 SMAD 蛋白发现及 TGF-b 信号转导研究的最激动人心的一年，哈佛医学院的 Whitman 实验室在十月份的《自然》杂志上以"Article"的形式发表了一篇名为 *A transcriptional partner for MAD proteins in TGF-b signaling* 的文章。读完之后，正好遇到 TGF-b 领域的著名学者 Joan Massague，我对 Joan 评论说：I'm not so sure why this paper deserves a full article in Nature. They just identified another Smad-interacting protein, and the data quality is mediocre. 完全出乎我的意料，Joan 马上回应我：I disagree! This paper links the cytoplasmic Smad protein into the nucleus and identifies a transcription factor as its interacting protein. Now the TGF-b signaling pathway is complete. It is a beautiful Nature article! 这件事对我触动极大：原来大师的视野和品位远远在我之上。从那以后，我也开始从整个领域的发展方面来权衡一篇文章的重要性，这件事对我今后为国际重要学术期刊审稿、自己实验室选择研究课题都起到了相当重要的作用。

如今，我阅读一篇本领域内的科研论文，非常顺利，而且常常可以看出一些作者没有想到或分析到的关键点。回想从前，感慨万千，感谢蒲慕明、郑昌学、虞一华、John Desjarlais、Jeremy Berg、Joan Massague 等一批老师和同事对我的帮助。我很留心，也很用心。

希望所有的学生也能通过努力和坚持对英文科研论文的阅读得心应手！

（http://blog.sciencenet.cn/blog-46212-350496.html）

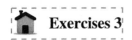

Exercises 3

Please search an academic article about ENVIRONMENT selected from *Science*, *Nature*, or *Scientific American*, then take notes about the article.

Basic Information about Your Drinking Water

By EPA Website on November 12, 2019

The United States enjoys one of the world's most reliable and safest supplies of drinking water. Congress passed the Safe Drinking Water Act (SDWA) in 1974 to protect public health, including by regulating public water systems.

The Safe Drinking Water Act (SDWA) requires EPA to establish and enforce standards that public drinking water systems must follow. EPA delegates primary enforcement responsibility (also called primacy) for public water systems to states and Indian Tribes if they meet certain requirements.

Approximately 150,000 public water systems provide drinking water to most Americans. Customers that are served by a public water system can contact their local water supplier and ask for information on contaminants in their drinking water, and are encouraged to request a copy of their Consumer Confidence Report. This report lists the levels of contaminants that have been detected in the water, including those by EPA, and whether the system meets state and EPA drinking water standards.

About 10 percent of people in the United States rely on water from private wells. Private wells are not regulated under the SDWA. People who use private wells need to take precautions to ensure their drinking water is safe.

（https://www.epa.gov/ground-water-and-drinking-water/basic-information-about-your-drinking-water）

Unit 3 Water Pollution and Control

Lesson 2 Conventional Wastewater Treatment Process

Part 1 Reading

1 Water Pollution

To understand the effects of water pollution and the technology applied in its control, it is useful to classify pollutant into various groups or categories. First, a pollutant can be classified according to the nature of its origin as either a point source or dispersed source pollutant.

A point source pollutant is one that reaches the water from a pipe, channel, or any other confined and localized source. The most common example of a point source of pollutants is a pipe that discharges sewage into a stream or river. Most of these discharges are treatment plant effluents, that is, treated sewage from a water pollution control facility: they still contain pollutants to some degree.

A dispersed or nonpoint source is a broad, unconfined area from which pollutants enter a body of water. Surface runoff from agricultural areas, for example, carries silt, fertilizers, pesticides, and animal wastes into streams, but not at one particular point. These materials can enter the water all along a stream as it flows through the area. Acidic runoff from mining areas is a dispersed pollutant. Storm water drainage systems in towns and cities are also considered to be dispersed sources of many pollutants because, even though the pollutants are often conveyed streams or lakes in drainage pipes or storm sewers, there are usually many of these discharge scattered over a large area.

Point source pollutants are easier to deal with than dispersed source pollutants; those from a point source have been collected and conveyed to a single point where they can be removed from the water in a treatment plant, and the point discharges from treatment plants can easily be monitored by regulatory agencies. Under the Clean Water Act, a discharge permit is required for all point sources.

Pollutants from dispersed sources are much more difficult to control. Many people think that sewage is the primary culprit in water pollution problems, but dispersed sources cause a significant fraction of the water pollution in the United States. The most effective way to control the dispersed sources is to put appropriate restrictions on land use.

In addition to being classified by their origin, water pollutants can be classified into groups of substances based primarily on their environmental or health effects. For example, the following list identifies nine specific types of pollutants:

1) Pathogenic organisms;
2) Oxygen-demanding substances;
3) Plant nutrients;
4) Toxic organics;
5) Inorganic chemicals;
6) Sediment;
7) Radioactive substances;
8) Heat;
9) Oil.

Domestic sewage is a primary source of the first three types of pollutants. Pathogens, or disease causing microorganisms, are excreted in the feces of infected persons and may be carried into waters receiving sewage discharges. Sewage from communities with large populations is very likely to contain pathogens of some type.

Sewage also carries oxygen-demanding substances—the organic wastes that exert a biochemical oxygen demand as they are decomposed by microbes. This is called BOD which changes the ecological balance in a body of water by depleting the dissolved oxygen (DO) content. Nitrogen and phosphorus, the major plant nutrients, are in sewage, too, as well as in runoff from farms and suburban lawns.

Conventional sewage treatment processes significantly reduce the amount of BOD in sewage, but do not eliminate them completely. Certain viruses, in particular, may be somewhat resistant to the sewage disinfection process. (A virus is an extremely small pathogenic organism that can only be seen with an electron microscope.) To decrease the amounts of nitrogen and phosphorus in sewage, usually some form of advanced sewage treatment must be applied.

Toxic organic chemicals, primarily pesticides, may be carried into water in the surface run off from agricultural areas. Perhaps the most dangerous type is the family of chemicals called chlorinated hydrocarbons. Common examples are known by their trade names as chlordane, dieldrin, heptachlor, and the infamous DDT, which has been banned in the United States[1]. They are very effective poisons against insects that damage agricultural crops. But, unfortunately, they can also kill fish, birds, and mammals, including humans. And they are not very biodegradable, taking more than 30 years in some cases to dissipate from the environment.

Toxic organic chemicals can also get into water directly from industrial activity, either from improper handling of the chemicals in the industrial plant or, as has been more common, from improper and illegal disposal of chemical wastes. Proper management of toxic and other hazardous wastes is a key environmental issue, particularly with respect to the protection of groundwater quality. Poisonous inorganic chemicals, specifically the heavy metal group such as lead, mercury and chromium, also usually originate from industrial activity and are considered hazardous wastes.

Oil is washed into surface waters in runoff from roads and parking lots, and ground water be polluted from leaking underground tanks. Accidental oil spills from large transport tankers at sea occasionally occur, causing significant environmental damage. And blowout accidents at offshore oil wells can release many thousands of tons of oil in a short period of time. Oil spills at sea may eventually move toward shore, affecting aquatic life and damaging recreation areas.

2 Waste Water Treatment

Waste-water treatment basically involves two sets of measure:

1) Preventive;

2) Curative.

The preventive steps are made up of

(i) Volume reductions of wastewater;

(ii) Strength reduction.

Volume reduction can be achieved by (a) conservation of water used in the process, (b) segregation of different streams in the process, (c) recycling and reusing water used in the process, (d) changing the production schedule to decrease waste water produced and (e) avoiding slug or batch discharges.

Strength reduction can be achieved by (a) equipment modifications and process changes, (b) segregation, equalization and proportioning and (c) recovery of important byproducts from waste streams.

The preventive measures outlined above are general and one or more of them may be applied, depending on the specific application. In a typical pharmaceutical industry, manufacturing drug intermediates, for example, it was found sufficient to segregate different waste streams and treat the most concentrated one for removal of dissolved inorganic chemical.

Curative measures deal with the actual treatment of liquid effluents by physical, chemical and biological methods, or their combinations, depending on the nature of the pollutants in the waste and the extent to which they are to be removed. The degree to which treatment is required depends upon the mode of disposal of the treated waste. Different standards have been laid down for the discharge of effluents into natural water bodies, municipal sewers, and into the land.

3 Physical Methods

Physical methods aim at removing solid or liquid pollutants, based on their density difference from water. They are essentially wastewater clarification methods and remove suspended or floating solids or liquids.

Evaporation of wastewater followed by incineration, crystallization or spay drying helps to recover the heat or materials dissolved in the original wastes. The phenol in coke-oven-plant effluents and certain pharmaceutical liquid wastes can be extracted by benzene, diisopropyl ether and other solvents[2]. The solvent can be separated from the extract layer by distillation or washing with alkalis. Other physical methods of treatment (referred to also as advanced waste-water-treatment methods) are reverse osmosis, electrodialysis, filtration, foam separation, porous-bed filtration, adsorption, etc. They help remove fine particles and organic and inorganic dissolved materials, resulting in better water quality for reuse or disposal[3]. Besides, some useful materials can also be recovered.

4 Biological Wastewater Treatment

In this method colloidal and dissolved solids are converted into settleable solids by micro-organisms under favourable environmental conditions, anaerobic treatment takes place in the total absence of oxygen and is a rather slow process. However, highly concentrated wastes can be handled by this method. Anaerobic filtration and digestion are some of its categories. Aerobic biological-treatment methods include the activated-sludge process, trickling-filter process and stabilization ponds.

In the activated-sludge process wastewater and flocculate-bacterial sludge are mixed and aerated in concrete, sealed earth basins or steel tanks. The bacteria grow at the cost of organic increases substantially. The design of an activated-sludge basin is determined by the biodegradability and the amount of organics present in the effluent. Aeration systems using compressed air make use of diffusers, injectors and static mixers. Recently oxygen instead of air is being increasingly employed. Surface aerators can be used for basin depths less than 4 m. Surface aerators can be easily repaired or replaced. A lot of mist, however, is produced in the vicinity of the treatment plants, causing undesirable odours when strong smelling wastes are processed.

Trickling filters consist of circular beds, 2-5 m high, filled with porous lumpy materials such as hard coke, slag or even stone-metal. The wastewater is evenly distributed on the surface of the bed with a rotary sprinkler. The slimy bacterial films formed on the surface of the packing take up organic materials from the wastewater as the latter trickles down the bed. The water is collected below the support plate or grating and is sent to a settling basin to remove the sludge. Plastic tubular or honeycomb-shaped packings can be used for highly-polluted wastes. There is lower energy consumption in the case of trickling filters as compared to activated-sludge processes and they are easier to maintain. On the other hand, capital cost is more and treatment capacity lower. Therefore, it is best to combine trickling filters with the activated-sludge process.

The stabilization ponds will be mentioned later.

5 Chemical Waste-Water Treatment

Industrial effluents usually contain acids, alkalis, chlorides, phenols, sulphates, chromates, phosphate, and salts of mercury, lead, calcium, barium, zinc, etc. The best way to remove these is to incorporate in plant changes in process design and operation. In fact, recovery and disposal of these pollutants should be incorporated in the process-design stage itself. The blowdown from the cooling tower contains chromium (usually used as a mixture with phosphate and zinc): the limit of hexavalent chromium for discharge into surface waters is 0.1mg/L. One scheme for the removal of chromium consists of acidification followed by reduction with ferrous sulphate and subsequent lime dosing.

Industrial waters usually contain acids and bases and need to be neutralized before they are discharged into water bodies or municipal sewers. Acids, when neutralized with caustic soda, give soluble salts. Caustic soda, however, is more expensive than lime which is, therefore, used for neutralization of acidic wastes. However, elaborate storage and grinding facilities, are required, the cost of which balances out the higher cost of caustic soda. When lime is used to neutralize hydrochloric acid in industrial waste-water, it gives soluble calcium chloride which is also undesirable and has to be treated further with soda (sodium carbonate) to give insoluble calcium carbonate. All this adds to the total sludge quantity. Neutralization of alkalis is done with a calculated quantity of sulphuric or hydrochloric acid. Tanks lined with acid-alkali-resistant material with mild agitation by similarly lined agitators, can be used for neutralization. A residence time of 10-20 min is considered sufficient for neutralization.

Flocculation and precipitation can help reduce the organic and inorganic load by settling the sludges or precipitates formed. More than 90 Percent of the dissolved phosphates and 55-70 percent of the BOD load of domestic sewage can be removed if domestic waste-water-clarification is combined with iron and aluminium salts. With clarification alone, only 25 percent of BOD can be removed. Dispersed polymers and oil emulsions can also be broken up by flocculation. Chromium (hexavalent) can be reduced to trivalent chromium with SO_2 gas, giving a precipitate of chromium sulphate.

Oxidation with chlorine and sodium hydrochlorate is used for killing pathogenic organisms in the treated liquid waste as well as for treating industrial wastes. Cyanides are oxidized to cyanate by chlorine or sodium hypochlorite[④]. Liquefied chlorine from cylinders is used in most applications.

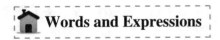

Words and Expressions

category	[ˈkætəɡɔːrɪ]	n. （人或事物的）类别，种类
facility	[fəˈsɪlətɪ]	n. 设施，设备，特别装置

scattered	[ˈskætərd]	adj. 分散的，零散的，疏落的
culprit	[ˈkʌlprɪt]	n. 罪犯，肇事者，引起问题的事物
pathogenic	[ˌpæθəˈdʒenɪk]	adj. 病原的，致病的
sediment	[ˈsedɪmənt]	n. 沉淀物，沉积物
decomposed	[ˌdiːkəmˈpoʊzd]	v. 腐烂，（使）分解；decompose 的过去分词和过去式
pesticide	[ˈpestɪsaɪd]	n. 杀虫剂，除害药物
nutrient	[ˈnuːtrɪənt]	n. 营养素，营养物
dieldrin	[ˈdɪldrɪn]	n. 狄氏剂，地特灵，氧桥氯甲桥萘
heptachlor	[ˈheptəˌklɔ]	n. 七氯
slug	[slʌg]	n. 少量，阻塞物
pharmaceutical	[ˌfɑːrməˈsuːtɪkl]	adj. 制药的，配药的
intermediate	[ˌɪntərˈmiːdɪət]	adj. （两地、两物、两种状态等）之间的，中间的，中级的，中等的
laid down		放下，铺设
incineration	[ɪnˈsɪnəˌreɪʃən]	n. 焚烧，焚化，煅烧，火葬，烧灼灭菌法，灰化作用
crystallization	[ˌkrɪstəlaɪˈzeɪʃn]	n. 结晶（作用，过程），结晶体，晶化，具体化，形象化
coke-oven-plant		n. 焦炉厂
diisopropyl ether	[daɪˌaɪsəˈproupil ˈiːθə]	n. 二异丙基醚，丙醚
solvent	[ˈsɑːlvənt]	n. 溶剂；adj. 有溶解力的，可溶解的
alkali	[ˈælkəˌlaɪ]	n. 碱
osmosis	[ɑːzˈmoʊsɪs]	n. 透析，渗透
electrodialysis	[ɪˌlektroʊdaɪˈæləsɪs]	n. 电渗析
foam separation		n. 泡沫分离法，泡沫分离
porous-bed filtration		n. 多孔滤床
digestion	[daɪˈdʒestʃən]	n. 厌氧消化
trickling-filter process		n. 滴滤法
flocculate-bacterial sludge		n. 絮凝菌泥
biodegradability	[baɪəʊdɪgreɪdəˈbɪlɪtɪ]	n. 可生化性，生物降解能力
injector	[ɪnˈdʒɛktər]	n. 注入极，喷射器（泵，头）喷嘴，注入（射、水、油）器，灌浆机
mist	[mɪst]	n. 薄雾，水汽，液体喷雾
lumpy	[trɪhələumeˈθeɪmz]	n. 三氯甲烷
coke	[koʊk]	n. 焦炭；

		v. （使）炭化
slag	[slæg]	*n.* 矿渣，熔渣，炉渣； *v.* 使成渣，使变成熔渣
slimy	[ˈslaɪmɪ]	*adj.* 泥浆样的
grating	[ˈgreɪtɪŋ]	*n.* （窗户、地沟口等的）栅栏，格栅
chloride	[ˈklɔːraɪd]	*n.* 氯化物
sulphate	[ˈsʌlfeɪt]	*n.* 矿物质，矿物
chromate	[ˈkrəumɪt]	*n.* 铬酸盐
phosphate	[ˈfɑːsfeɪt]	*n.* 磷酸盐，含磷化合物，磷肥
lead	[lɪːd]	*n.* 铅
barium	[ˈberɪəm]	*n.* 钡
blowdown	[ˈbləudaun]	*n.* 吹脱，排污，吹除，放气，（发动机试验后）吹净
hexavalent	[heksəˈveɪlənt]	*adj.* 六价的
acidification	[əˌsɪdəfəˈkeɪʃən]	*n.* 酸化，成酸性，酸化作用
ferrous	[ˈferəs]	*adj.* 含铁的，铁的
base	[beɪs]	*n.* 根基，源泉，基础
caustic soda	[ˌkɔːstɪk ˈsoudə]	*n.* 苛性钠，烧碱，氢氧化钠
acidic	[əˈsɪdɪk]	*adj.* 酸性的
elaborate	[ɪˈlæbərət]	*adj.* 复杂的，详尽的，精心制作的； *v.* 详尽阐述，详细制订，精心制作
balance out		*v.* （使）平衡，相抵，抵消
hydrochloric acid		*n.* 盐酸
calcium chloride		*n.* 氯化钙
sodium carbonate		*n.* 碳酸钠
calcium carbonate		*n.* 碳酸钙
sulphuric	[sʌlˈfjʊərɪk]	*n.* 硫酸，硫黄，硫的
polymer	[ˈpɑːlɪmər]	*n.* 聚合物，多聚体

Notes

① Perhaps the most dangerous type is the family of chemicals called chlorinated hydrocarbons. Common examples are known by their trade names as chlordane, dieldrin, heptachlor, and the infamous DDT, which has been banned in the United States.

参考译文：也许最危险的是一系列被称为氯代烃的化学物质。它们的商品名更为常见，如氯丹、狄氏剂、七氯以及臭名昭著的，已在美国被禁止使用的 DDT。

② The phenol in coke-oven-plant effluents and certain pharmaceutical liquid wastes can be extracted by benzene, diisopropyl ether and other solvents.

参考译文：焦炉废水和某些制药废液中的苯酚可用苯、二异丙基醚等溶剂萃取。

③ Other physical methods of treatment (referred to also as advanced waste-water-treatment methods) are reverse osmosis, electrodialysis, filtration, foam separation, porous-bed filtration, adsorption, etc. They help remove fine particles and organic and inorganic dissolved materials, resulting in better water quality for reuse or disposal.

参考译文：其他物理处理方法（也称为高级废水处理方法）有反渗透、电渗析、过滤、泡沫分离、多孔床过滤、吸附等，它们有助于去除细小颗粒和有机、无机溶解物，使水质更好，以便回用或处置。

④ Cyanides are oxidized to cyanate by chlorine or sodium hypochlorite.

参考译文：氰化物可以被氯气或次氯酸钠氧化成氰酸盐。

Exercises 1

1. According to the reading material, chose the best answer(s) from the options.

(1) Water pollutant can be classified according to the nature of its origin as _____ and _____.

  A. point source   B. dispersed source.

  C. outdoor source   D. domestic source

(2) The most effective way to control the dispersed sources is _____.

  A. abandon agricultural practices   B. to treat the pipe discharge

  C. to put appropriate restrictions on land use   D. to collect the wastewater

(3) Domestic sewage is a primary source of _____, _____ and _____.

  A. Pathogenic organisms   B. Oxygen-demanding substances

  C. Plant nutrients   D. Toxic organics

(4) Conventional sewage treatment processes significantly reduce the amount of _____ in sewage.

  A. nitrogen   B. virus   C. DO   D. BOD

(5) Curative measures deal with the actual treatment of liquid effluents by _____, _____ and _____ methods.

  A. physical   B. biological   C. chemical   D. advanced

(6) Biological treatment includes _____ and _____.

  A. anaerobic treatment   B. filtration

  C. aerobic biological-treatment   D. foam separation

2. Propose one method of biological wastewater treatment.

3. Propose one method of physical wastewater treatment.

4. Propose one method of chemical wastewater treatment.

Part 2　Translation

英语翻译中的增译

英汉词汇在意义范围、使用习惯、修辞等方面不尽相同，在翻译时，如果追求对等译文则往往会形神偏离。因此，增词法应运而生，作为常见的有效补偿手法，也称为增译。增译就是在原文的基础上，在翻译时按照意义、修辞或句法上的需要添加必要的单词、词组或句子，从而使原文在语法和语言形式上符合译文的语言习惯和文化背景。

一、语法需要引起的增译

1. 增加表示复数概念的词/词组

一般来说，英语中表示复数的词可以在名词后面直接加"s"，或单复数同行，一些不规则的词则有特殊的形式，在形式上体现明显。

而汉语则多用一些词语来表达复数的概念，如"许多""大量""种种""若干""排排""朵朵""层层"等，例如：

(1) Perhaps the most dangerous type is the family of chemicals called chlorinated hydrocarbons.

译文：也许最危险的是一系列被称为氯代烃的化学物质。

(2) Radioactive materials can be used to learn how the glands and organs of our bodies function.

译文：放射性物质可用来查明各种腺体和器官的功能。

(3) Pictures taken from the space show the changes that both natural and human forces have brought about on the earth's surface.

译文：在太空拍摄的照片展示出大自然和人类的力量使地球表面发生的种种变化。

2. 增加表示数量的数词、量词

科技英语的一个重要特征是表达科学、准确，来不得半点模糊，那么译文也势必要准确。

例如：

(1) The first electronic computers used vacuum tubes and other components, and this made the equipment very large and bulky.

译文：第一批电子计算机使用真空管和其他元件，这使得设备又大又笨重。

(2) If you press a spring axially with your hands at its ends, you will feel a reaction from it.

译文：如果你用双手从两端压一根弹簧，你就会感到弹簧有一种反作用力。

(3) The revolution of the earth around the sun causes the changes of the seasons.

译文：地球围绕太阳旋转，引起四季的交替。

3. 省略词的增译

由于英语具有避开使用重复词，保持语言结构简洁的习惯，所以往往会省略一些重复词和无足轻重的词语。翻译时，为了使译文符合原文和汉语表达，我们需要补全这些省略的词。例如：

(1) The best conductor has the least resistance and the poorest the greatest.

译文：最好的导体电阻最小，最差的导体电阻最大。（增加了省略的"导体电阻"一词）

(2) This year the production of TV sets in our plant has increased by 20 percent, and of tape recorder by 15 percent.

译文：今年我们厂的电视机产量增长了20%，磁带录音机产量增长了15%。（增加了"增长"一词）

(3) Care should be taken at all times to protect the instrument from the dust and damp.

译文：应经常注意保护仪器，勿使其沾染灰尘，勿使其受潮。（根据句子内涵意思，增加了"勿使其沾染""勿使其受潮"）

4. 非谓语动词、分词增译

英语和汉语不同，英语句式多样，各种句式都体现一个意思，非谓语动词、分词短语是英语中常见句式，在独立结构中用作状语，用来修饰主句中的谓语动词或主句，表示暗含的时间、原因、条件、方式、结果、让步或伴随意义，在翻译的时候，必须通过增译，准确译出这些非谓语动词、分词短语的确切意义。例如：

(1) Possessing high conductivity of heat and electricity, aluminum finds wide application in industry.

译文：铝具有良好的导热性和导电性，故在工业上得到广泛的应用。（分词表示"原因"）

(2) Examining a drop of water under the microscope, we shall see a huge number of bacteria.

译文：如果把一滴水放在显微镜下观察，就会看到大量的细菌。（分词表示"假设"）

5. 被动语态的增译

英语与汉语之间的一大差别就是英语喜欢使用被动语态表示主动的意思，一些描述客观事实的科技材料，也大量使用被动句式，在翻译过程中，译者必须将其译为符合汉语习惯的主动态，例如：

(1) A comprehensive review is made covering various aspects of the application d

译文：本文旨在全面论述激光的应用。（增加了主语"本文"）

(2) It is believed to be natural that more and more engineers have come to prefer synthetic material to natural material.

译文：越来越多的工程人员宁愿用合成材料而不用天然材料，人们相信这是很自然的。（增加了"人们"一词）

6. 时态的增译

英语动词的时态种类繁多，往往是靠自身形态的变化或增加助动词加以实现，而汉语则不同，汉语没有词性的变化，时间概念可以体现在字里行间，或是通过借用一般的时态助词、表时间概念的副词，如"能""会""曾经""现在""将"等来实现，如：

(1) Research has shown a connection between smoking and cancer.

译文：研究已经证明抽烟和癌症间的关系。（增加了"已经"一词）

(2) The phenol in coke-oven-plant effluents and certain pharmaceutical liquid wastes can be extracted by benzene, diisopropyl ether and other solvents.

译文：焦炉废水和某些制药废液中的苯酚可用苯、二异丙基醚等溶剂萃取。（增加了"可用"一词）

7. 动词语气的增译

英语动词可以表达陈述、祈使和虚拟三种语气。汉语语气的表达一般不需要加词，但需要增加词语才能反映出来，才能加强表达的气势或使译文更加通畅、可读。如：

Inspect the instrument carefully for damage. Should the instrument show any signs of damage, file a claim with the carrier immediately.

译文：请仔细检查仪器有无损坏现象。如果损坏，应立即向承运者提出索赔要求。（增加了"请""如果"两词）

二、意义表达引起的增译

由于英汉语言的差异，翻译中常常会发现原文中包含一部分省略或隐含的词义，而这些语义必须在译文中充分表达出来。在遇到诸如科技英语等专业性较强的文本，其语言结构偏向简练，在汉译的过程中，需要仔细体会，适当增加合适的词，一般来说，既可以增加词语的含义，也可以增加范畴词，如方法、现象、效应、作用、状态等，被增加的范畴名词大多表示抽象概念，这样既可使译文流畅，又能被读者接受。例如：

1. 名词的增译

前面词义的引申一节中讲过了具体化引申，英语有很多抽象化的名词，是由形容词或动词派生而来，表示动作或行为意义，在翻译过程中，要注意英语这样一个用词特点，将抽象名词具体化，或根据具体的汉语习惯搭配增加一些词/词组。例如：

(1) This transfer continues until a uniform temperature is reached, at which point no further energy transfer is possible.

译文：这一传递过程一直延续到温度均匀时为止，在温度均匀时能量便不可能进一步传递。（增加了"过程"一词）

(2) Electricity permits indication of measurements at a distance, which is very important in many fields.

译文：电可以在相隔一段距离的地方显示各种测量结果，这在许多场合都是很重要的。（增加了"结果"一词）

2. 动词的增译

根据意思需要，在名词（有的是由动词演变而来的动名词）前加上动词，是英语中常用的一种技巧，如 programming（编写程序），instrumentation（研制仪器、使用仪器）。又如：

(1) Testing is a complicated problem, so be careful.

译文：进行测试是一个复杂的问题，因此，要格外仔细。（增加了动词"进行"一词）

(2) It is not surprising, then, that the world saw a return to a floating exchange rate system. Central banks were no longer required to support their own currencies.

译文：在这种情况下，世界各国恢复采用浮动汇率就不足为奇了。中央银行就无需维持本币的汇价了。（增加了"采用"二字）

3. 形容词的增译

在名词前加上必要的形容词，但这个形容词必须是这个名词所蕴含的，可以使表达更为准确，例如：

(1) The agency soon moved him from forgery to disguises like mustaches and wings, which were fairly crude back then in the 1960s.

译文：很快，中央情报局不再让他造假，而改搞化妆，如贴假胡子、戴假发，而在20世纪60年代这些东西制作得相当粗糙。（增加了形容词"假的"）

此外，为了意义的明确表达，还可以重复同一词语、增补连词或承上启下的词等。如：

(2) Energy may or not be the primary factor in system design.

译文：在系统设计中，能源可能是第一要素，也可能不是第一要素。（增加了重复词"第一要素"）

(3) How these two things—energy and matter-behave, how they interact one with the other, and how people control them to serve themselves make up the substance of two basic physical sciences, physics and chemistry.

译文：能量和物质这两样东西具有什么性质，如何相互起作用，如何控制它们为人类自身服务，这些问题就构成了物理和化学这两门基础自然科学的主要内容。（增加了承上启下的"这些问题"）

三、逻辑需要引起的增译

由于英汉两种语言的表达句式不同,有些英语句子在逻辑上是通的,但如果按照原句"直译"过来,在汉语中就会显得极其怪异。因此,在翻译时,需要加上适当的词或短语确保其在逻辑上流畅,但又和原文意思相吻合。增加必要的逻辑词(也包括一些承上启下的词、概括性的词等)需要译者有一定的逻辑推理能力,将原文中隐含的意义表达出来。例如:

(1) Since air has weight, it exerts force on any object immersed in it.

译文:因为空气具有重量,所以处在空气中的任一物体都会受到空气的作用力。(根据逻辑需要增加了"所以")

(2) For example, if a factory closes down because it cannot meet government pollution standards, a large number of workers suddenly find themselves without jobs.

译文:假如一家工厂因为在防止污染方面达不到政府所规定的标准而关门,相当多的工人就会突然失业。(增加了逻辑上的"防止"一词)

(3) The concentration of carbon dioxide in the air being only 0.03 percent, carbon is amassed into the compass of the plant from a large volume of air.

译文:因为空气中二氧化碳的密度只有万分之三,所以聚集在植物内部的碳是从周围大量的空气中得来的。(增加了因果关系词"因为""所以")

四、修辞需要引起的增译

英汉语言之间的巨大差异还体现在其修辞上。为了把原文的意思表达清楚,使译文更加优美、准确、生动、鲜明,需要对译文进行必要的修辞上的润色,也可以增加适当的修饰词,如名词、动词、形容词、副词、语气助词等。总之,修辞引起的增译范围比较广泛,这需要译者具备一定的审美意识。例如:

(1) But the appearance of less vulnerability of America to supply interruption of imported oil is deceptive and dangerous.

译文:乍一看,进口石油供应中断已不再会对美国造成严重损害了,但这种印象具有欺骗性和危险性。(增加了语气助词"乍一看")

(2) When the aircraft was towed to the terminal building, the crew stood at the top of the stairs to shake hands with passengers as they disembarked.

译文:在飞机被牵引到候机大楼外时,全体机组人员都站在楼梯口与走下飞机的所有乘客一一握手。(增加了"一一")

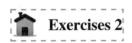
Exercises 2

Translate the following paragraph into Chinese.

(1) Many people think that sewage is the primary culprit in water pollution problems, but

dispersed sources cause a significant fraction of the water pollution in the United States. The most effective way to control the dispersed sources is to put appropriate restrictions on land use.

(2) Flocculation and precipitation can help reduce the organic and inorganic load by settling the sludges or precipitates formed. More than 90 Percent of the dissolved phosphates and 55-70 percent of the BOD load of domestic sewage can be removed if domestic aste-water-clarification is combined with iron and aluminium salts. With clarification alone, only 25 percent of BOD can be removed. Dispersed polymers and oil emulsions can also be broken up by flocculation. Chromium (hexavalent) can be reduced to trivalent chromium with SO_2 gas, giving a precipitate of chromium sulphate.

Part 3　Writing

学术论文的撰写（三）——学术论文提纲的写作技巧

在写作中，提纲有利于把写作的内容按照一种逻辑性强、思路清楚且有效的方式展开。提纲不仅能表示文章的主题思想，而且能表明文章的结构和说明各个概念之间的关系。

在提纲中，第一段通常表示的是控制性的主题。其他部分则根据这一主题进行扩展。为了写好提纲，可先将与主题有关的内容逐一列出，然后把这些内容按层次组织成相等的部分，最后再将这些内容具体化。提纲写作的方法主要有两种，一种是条目性提纲（topic outline），另一种是句子性提纲（sentence outline）。

一、条目性提纲

条目性提纲是用名词词组或相当于名词词组的组成，见下列实例：

【例1】Topic: Water Pollution

Thesis statement: More should be done to control water pollution

Outline:　Ⅰ. A brief introduction to the problem of water pollution

Ⅱ. Main causes of water pollution

1. pollutants from factories

2. remains from insects killers

3. wastes from people

Ⅲ. Results of water pollution

1. harm to health

2. bad influence on the environment

3. serious effect on future development

Ⅳ. Some suggested measure

1. government

2. factories

3. individuals

Ⅴ. Conclusion

二、句子性提纲

句子性提纲是使用一个完整的句子来表示提纲的内容。见下列实例：

【例2】Topic: Travelling

Thesis statement: Travelling can enrich one's experience

Outline:　Ⅰ. Introduction: Travelling can enrich one's experience.

Ⅱ. Travelling involves various activities.

1. One makes the plan including scheduling, budgeting and ticketing.

2. One needs to accommodate oneself.

Ⅲ. Travelling offers one an opportunity to make contacts with different people.

1. On the way, one can communicate with other passengers.

2. At the destination, one can communicate with the local people.

Ⅳ. Travelling gives one an opportunity to see different places.

1. One can learn from historic places.

2. One can enjoy the beauty of natural scenery.

3. One can have a chance to learn about the local culture

Ⅴ. Conclusion: by traveling, one will become more knowledgeable and experienced.

通过对以上两个实例的对比，可以发现条目性提纲比句子性提纲要简洁，而后者则有利于段落的展开。

在实践中，比较实用的提纲是非正式的、草稿性的提纲，这种提纲只要能启发写作就行了，因为它可以使作者按照所确定的思路进行扩展。在起草提纲时，还可根据写作目的列出一些词汇和短语，然后对其进行调整，以便于在写作框架中能体现出文章的连贯性和完整性。这样，文章的结构以及对主题进行扩展的脉络就更清楚了。

需要说明的是，在草拟提纲时，要清楚提纲仅仅是个思路，在写作时不需要绝对按其要求来完成论文。实际上，在写作中提纲的内容也是会常常修改的。一般来说，经过修改的提纲有如下特点：

（1）克服了惰性。只要有了草拟的标题就有了一个良好的开端，它有利于排列出条目，有助于使思想内容具体化。

（2）重点呈现需要仔细讨论的要点。在撰写论文的过程中，常常会在细节描述时发生偏离，对要点列出提纲则可避免这种情况。如果事先对需要仔细探讨的要点排列了顺序，在写作时可以参照提纲的顺序写作，这样，就不会偏离要点或忽略某一重要的问题。

（3）可作为文章要点的组织者以及对要点进行讨论的引导者。

如果作者是将重点放在功能上，而不是形式上，提纲就是一个非常有用的工具，它可以节省实际写作的时间。

 **Exercises 3**

Please search an academic article about WASTE WATER TREATMENT selected from *Science*, *Nature*, or *Scientific American*, then summarize the outline of the article.

 **Expanding Reading**

Municipal Wastewater

Overview

The collection and treatment of domestic sewage and wastewater is vital to public health and clean water. It is among the most important factors responsible for the general level of good health enjoyed in the United States. Sewers collect sewage and wastewater from homes, businesses, and industries and deliver it to wastewater treatment facilities before it is discharged to water bodies or land, or reused.

Wastewater Treatment Facilities

NPDES permits establish discharge limits and conditions for discharges from municipal wastewater treatment facilities to waters of the United States. Resources for discharge requirements include:

- Primer for Municipal Wastewater Treatment
- NPDES (National Pollutant Discharge Elimination System) Permitting Framework
- Secondary Treatment Standards

Collection Systems

Historically, municipalities have used two major types of sewer systems.

- Combined Sewers

Combined sewers are designed to collect both sanitary sewage and storm water runoff in a single-pipe system. These systems were designed to convey sewage and wastewater to a treatment plant during dry weather. Under wet weather conditions, these combined sewer systems would overflow during wet weather conditions when large amounts of storm water would enter the system. State and local authorities generally have not allowed the construction of new combined sewers since the first half of the 20th century.

- Separate Sanitary Sewers

The other major type of domestic sewer design is sanitary sewers (also known as separate sanitary sewers). Sanitary sewers are installed to collect wastewater only and do not provide widespread drainage for the large amounts of runoff from precipitation events. Sanitary sewers are typically built with some allowance for higher flows that occur when excess water enters the collection system during storm events.

Sanitary sewers that are not watertight due to cracks, faulty seals, and/or improper connections can receive large amounts of infiltration and inflow (I/I) during wet weather. Large volumes of I/I can cause sanitary sewer overflows (SSOs) and/or operational problems at the wastewater treatment facility serving the collection system. In addition, sewage overflows can be caused by other problems such as blockages, equipment failures, broken pipes, or vandalism.

(*https://www.epa.gov/npdes/municipal-wastewater*)

Unit 3 Water Pollution and Control

Lesson 3 Advanced Wastewater Treatment

Part 1 Reading

Secondary treatment can remove between 85 and 95 percent of the BOD and TSS in raw sanitary sewage. Generally, this leaves 30 mg/L or less of BOD and TSS in the secondary effluent. But sometimes this level of sewage treatment is not sufficient to protect the aquatic environment. For example, periodic low flow rates in a trout stream may not provide the amount of dilution of the effluent that is needed to maintain the necessary DO levels for trout survival.

Another limitation of secondary treatment is that it does not significantly reduce the effluent concentrations of nitrogen and phosphorus in the sewage. Nitrogen and phosphorus are important plant nutrients. If they are discharged into a lake, algal blooms and accelerated lake aging or eutrophication may be the result. Also, the nitrogen in the sewage effluent may be present mostly in the form of ammonia compounds. These compounds are toxic to fish if the concentrations are high enough. Yet another problem with the ammonia is that it exerts a nitrogenous oxygen demand in the receiving water, as it is converted to nitrates. This process is called nitrification.

When pollutant removal greater than that provided by secondary treatment is required, either further reduce the BOD or TSS concentrations in the effluent or to remove plant nutrients. Additional or advanced treatment steps are required. This is also called tertiary treatment, because any of the additional processes follow the primary and secondary processes in sequence.

Tertiary treatment of sewage can remove more than 99 percent of the pollutants from raw sewage and can produce an effluent of almost drinking water quality. But the cost of tertiary treatment, for operation and maintenance as well as for construction, is very high, sometimes doubling the cost of secondary treatment. The benefit-to-cost ratio is not always big enough to justify the additional expense. Nevertheless, application of some form of tertiary treatment is not uncommon. Some of the more common tertiary processes are discussed in the following sections.

1 Effluent Polishing

The removal of additional BOD and TSS from secondary effluents is sometimes referred to as effluent polishing. It is most often accomplished using a granular-media filter, much like the filters used to purify drinking water. Since the suspended solids consist mostly of organic compounds, filtration removes BOD as well as TSS.

Generally, mixed-media filters are used to achieve in-depth filtration of the effluent. Because of the organic and biodegradable nature of the suspended solids in the secondary effluent, tertiary filters must be backwashed frequently. Otherwise, decomposition would cause septic or anaerobic conditions to develop in the filter bed. In addition to the conventional backwash cycle, an auxiliary surface air-wash is used to thoroughly scour and clean the filter bed. Filtration may be done by gravity in an open tank or by pressure in closed pressure vessels.

The filtered water may be stored in an adjacent tank and used for back wash water when the head loss through the automatic-back wash tertiary filter reaches a predetermined level[①]. Some filter manufacturers mount the backwash storage tank directly above the filter, forming a single self-contained.

Another process, called microstraining, also finds application as a tertiary step in wastewater treatment for suspended solids reduction. The microstrainers, also called microscreen, are composed of specially woven steel wire cloth mounted around the perimeter of a large revolving drum. The steel wire cloth acts as a fine screen, with openings as small as 20 micrometers（μm, or millionths of a meter）.

The rotating drum is partially submerged in the secondary effluent, which must flow into the drum and then outward through the microscreen. As the drum rotates, captured solids are carried to the top, where a high-velocity water spray flushes them into a hopper mounted on the hollow axle of the drum[②].

2 Phosphorus Removal

Phosphorus is one of the plant nutrients that contributes to the eutrophication of lakes. Raw sewage contains about 10 mg/L phosphorus, from household detergents as well as from sanitary wastes. The phosphorus in wastewater is primarily in the form of organic phosphorus and as phosphates, PO_4^{3-}, compounds. Only about 30 percent of this phosphorus is removed by the bacteria in a conventional secondary sewage treatment plant, leaving about 7 mg/L of phosphorus in the effluent.

When stream or effluent standards require lower phosphorus concentrations, a tertiary treatment process must be added to the treatment plant. This usually involves and coagulation. The organic phosphorus compounds are entrapped in the coagulant flocs that are formed and settle out in a clarifier.

One chemical frequently used in this process is aluminum sulfate, $Al_2(SO_4)_3$. This is called alum, the same coagulant chemical used to purify drinking water. The aluminum ions in the alum react with the phosphate ions in the sewage to form the insoluble precipitate called aluminum phosphate. Other chemicals that may be used to precipitate the phosphorus include ferric chloride, $FeCl_3$, and lime, CaO.

Adding the coagulant downstream of the secondary processes provides the greatest overall reliability for phosphorus reduction. It not only removes about 90 percent of the phosphorus, but in removes additional TSS and serves to polish the effluent as well. But when applied in this manner, as a third or tertiary treatment step, additional flocculation and settling tanks must be built. In some cases, even filters may have to be added to remove the nonsettleable floc.

To avoid the need for construction of additional tanks and filters, in most plants requiring phosphorus removal the coagulant is added to the wastewater at some point in the conventional process. For example, alum may be added just before the primary settling tanks. The resulting combination of primary and chemical sludge is removed from the primary clarifiers.

Or, in activated sludge plants, the coagulant may be added directly into the aeration tanks. In this case, the precipitation and flocculation reactions occur along with the biochemical reactions. Sometimes, the coagulant may be added to the wastewater just before the secondary or final clarifiers. Regardless of the point in the process at which coagulant is added, the total volume and weight of sludge requiring disposal increase significantly.

3   Nitrogen Removal

Nitrogen can exist in wastewater in the form of organic nitrogen, ammonia, or nitrate compounds. The effluent from a conventional activated sludge plant contains mostly the ammonia nitrogen form, NH_4^+. Effluents from a trickling filter or rotating biodisc may contain more of the nitrate form, NO_3^-. This is because the nitrifying bacteria, those microbes that convert ammonia to nitrate, have a chance to grow and multiply on some of the surfaces in the trickling filter or biodisc units. They do not survive in a mixed-growth aeration tank, where they are crowded out by faster-growing bacteria that consume carbonaceous organics.

Nitrogen in the form of ammonia can be toxic to fish, and it exerts an oxygen demand in receiving waters as it is converted to nitrate, Nitrate nitrogen is one of the major nutrients that causes algal blooms and eutrophication. For these reasons, it is sometimes necessary to remove the nitrogen from the sewage effluent before discharge. This is particularly important if it is discharged directly into a lake.

One of the methods used to remove nitrogen is called biological nitrification-denitrification. It consists of two basic steps. First, the secondary effluent is introduced into another aeration tank, trickling filter, or biodisc. Since most of the carbonaceous BOD has already been removed, the microorganisms that will now thrive in this tertiary step are the nitrifying bacteria, Nitrosomonas and Nitrobacter. In this first step, called nitrification, the ammonia nitrogen is converted to nitrate nitrogen, producing a nitrified effluent. At this point, the nitrogen has not actually been removed, but only converted to a form that is not toxic to fish and that does not cause an additional oxygen demand.

A second biological treatment step is necessary to actually remove the nitrogen from the wastewater. This is called denitrification. It is an aerobic process in which the organic chemical methanol is added to the nitrified effluent to serve as a source of carbon. The denitrifying bacteria Pseudomonas and other groups use the carbon from the methanol and the oxygen from the nitrates in their metabolic processes. One product of this biochemical reaction is molecular nitrogen, N_2, which escapes into the atmosphere as a gas.

Another method for nitrogen removal is called ammonia stripping. It is a physical-chemical rather than a biological process, consisting of two basic steps. First, the pH of the wastewater is raised in order to convert the ammonium ions, NH_4^+, to ammonia gas, NH_3. Second, the wastewater is cascaded down through a tall tower; this causes turbulence and contact with air allowing the ammonia to escape as a gas. Large volumes of air are circulated through the tower to carry the gas out of the system. The combination of ammonia stripping with phosphorus removal using lime as a coagulant is advantageous, since the lime can also serve to raise the pH of wastewater. Ammonia stripping is less expensive than biological nitrification-denitrification, but it does not work very efficiently under cold weather conditions.

4 Land Treatment of Wastewater

The application of secondary effluent onto the land surface can provide an effective alternative to the expensive and complicated advanced treatment methods previously discussed. A high-quality polished effluent can be obtained by the natural processes that occur as the effluent flows over the vegetated ground surface and percolates through the soil.

An additional benefit of land treatment is that it can provide the moisture and nutrients needed for vegetation growth, and it can help to recharge groundwater aquifers. In effect, land treatment of wastewater allows a direct recycling of water and nutrients for beneficial use: the sewage becomes a valuable natural resource that is not simply disposed of[③]. But relatively large land areas are needed for this land of treatment. And soil types as well as climate are critical factors in controlling the feasibility and design of a land treatment process.

There are three basic types or modes of land treatment: slow rate, rapid infiltration, and overland flow. The conditions under which they can function and the basic objectives of these types of treatment vary.

In the slow rate system, also called irrigation, vegetation is the critical component for the wastewater treatment process. Although the basic objective is wastewater treatment and disposal, another goal is to obtain an economic benefit from the use of the water and nutrients to produce marketable crops (that is, corn or grain) for animal feed. Another objective might be to conserve potable water by using secondary effluent to irrigate lawns and other landscaped areas.

The rapid infiltration or infiltration-percolation mode of land treatment has as basic objectives to recharge groundwater aquifers and to provide advanced treatment of wastewater. Most of the secondary effluent percolates to the groundwater: very little of it is absorbed by vegetation. The filtering and adsorption action of the soil removes most of the BOD, TSS, and phosphorus from the effluent, but nitrogen removal is relatively poor. Soils must be highly permeable for the rapid infiltration method to work properly. Usually, the wastewater is applied in large ponds called recharge basins.

In an overland flow system, wastewater is sprayed on a sloped terrace and allowed to flow across the vegetated surface to a runoff collection ditch[④]. Purification is accomplished by physical, chemical and biological processes as the wastewater flows in a thin film down the relatively permeable surface. Overland flow can be used to achieve removal efficiencies for BOD and nitrogen comparable to other methods of tertiary treatment, but phosphorus removal is somewhat limited. The water collected in the ditch is usually discharged to a nearby body of surface water.

Words and Expressions

sequence	[ˈsiːkwəns]	n. 一系列，顺序，次序； v. 按顺序排列，测定（整套基因或分子成分的）序列
maintenance	[ˈmeɪntənəns]	n. 维护，保养，维持，保持
decomposition	[ˌdiːˌkɑːmpəˈzɪʃn]	n. 腐烂，分解
adjacent	[əˈdʒeɪsnt]	adj. 与……毗连的，邻近的
feasibility	[ˌfiːzəˈbɪləti]	n. 可行性，可能性
eutrophication	[juːtrəfɪˈkeɪʃn]	n. （由雨水带来的化肥等造成水体的）富营养化
granular-media		颗粒介质
scour	[ˈskaʊər]	v. （彻底地）搜寻，翻找，（用粗糙的物体）擦净，擦亮，冲刷成，冲刷出
microstraining	[maɪkrɒstˈreɪnɪŋ]	n. 微过滤
flocculation	[ˌflɒkjʊˈleɪʃən]	n. 絮凝，絮结产物
carbonaceous	[ˌkɑːbəˈneɪʃəs]	adj. 含碳的，碳质的
nitrate	[ˈnaɪtreɪt]	n. 硝酸盐，硝酸盐类化肥； vt. 用硝酸处理，使硝化
nitrifying	[ˈnaɪtrɪfaɪɪŋ]	v. （使）硝化

denitrification	[dɪˌnaɪtrɪfɪˈkeɪʃn]	n. 反硝化作用，脱硝作用，脱氮作用
Nitrosomonas	[naɪtroʊˈsɒmənəs]	n. 亚硝酸菌属，亚硝化单胞菌属
pseudomonas	[ˌsudəˈmoʊnəz]	n. 假单胞菌属
percolate	[ˈpɜːrkəleɪt]	v. 渗入，渗透，渗漏，逐渐流传
infiltration-percolation		n. 渗滤
permeable	[ˈpɜːrmɪəbl]	adj. 可渗透的，可渗入的

Notes

① The filtered water may be stored in an adjacent tank and used for back wash water when the head loss through the automatic-back wash tertiary filter reaches a predetermined level.

参考译文：过滤后的水可以存储在相邻的水箱中，当通过自动反冲洗三级过滤器的水头损失达到预定水平时，可用作反冲洗水。

② The rotating drum is partially submerged in the secondary effluent, which must flow into the drum and then outward through the microscreen. As the drum rotates, captured solids are carried to the top, where a high-velocity water spray flushes them into a hopper mounted on the hollow axle of the drum.

参考译文：转筒部分在二级出水口以下，二级出水必须流入转筒，然后通过微孔筛网向外流出。当转筒旋转时，捕获的固体被带到顶部，再被高速喷水冲到转筒空心轴上的漏斗中。

③ In effect, land treatment of wastewater allows a direct recycling of water and nutrients for beneficial use: the sewage becomes a valuable natural resource that is not simply disposed of.

参考译文：事实上，废水的土地处理技术可以直接循环利用水和养分，使其得到有益的利用：污水成为一种宝贵的自然资源，而不是简单地加以处理。

④ In an overland flow system, wastewater is sprayed on a sloped terrace and allowed to flow across the vegetated surface to a runoff collection ditch.

参考译文：在地表径流系统中，废水被喷洒在坡式梯田上，并通过遍布植被的坡面流向截流沟。

Exercises 1

1. According to the reading material, chose the best answer(s) from the options.

(1) Another limitation of secondary treatment is that it does not significantly reduce the effluent concentrations of _____ and _____ in the sewage.

    A. BOD   B. DO   C. nitrogen   D. phosphorus

(2) Tertiary treatment of sewage can remove more than _____ percent of the pollutants from raw sewage and can produce an effluent of almost drinking water quality.

A. 80	B. 85	C. 90	D. 99

(3) Chemicals that may be used to precipitate the phosphorus include _____ , _____ and _____ .

A. $Al_2(SO_4)_3$	B. $FeCl_3$	C. CaO	D. $AlPO_4$

(4) One of the methods used to remove nitrogen is called biological nitrification-denitrification. It consists of _____ basic steps.

A. 1	B. 2	C. 3	D. 4

(5) Ammonia stripping and biological nitrification-denitrification, which is more expensive_____?

A. Ammonia stripping		B. Biological nitrification-denitrification

C. They are the same.		D. They can't be compared.

(6) There are three basic types or modes of land treatment: _____ , _____ and _____ .

A. slow rate		B. filtration

C. rapid infiltration		D. overland flow

2. Propose the methods of advanced wastewater treatment.

Part 2　Translation

英语翻译中的减译

在翻译中，增译和减译是一对相辅相成的翻译技巧概念。一般而言，英语省略的，汉语译文便应当增补；若英语繁复，则意味着汉语译文中应该有所取舍。减译，也称作省略译法或减词译法，就是在译文语法和修辞的基础上把原文中需要而译文中不需要的词、词组等在翻译过程中加以省略，使之符合目的语表达习惯。在科技翻译过程中，译者需要敏锐的目光，该减则减，不必手软，以确保译文简洁明快、严谨精练。以下就从语法需要、意义需要和修辞逻辑需要三个方面的减译来加以探讨。

一、语法需要引起的减译

由语法需要进行的减译，有时也称作虚词的减译。由于英语属于印欧语系，而汉语属于汉藏语系，因此在句法结构上存在巨大的差异。英语具有极其严密的语法系统，而汉语相对来说语法系统较弱，在翻译过程中，可以根据具体情况加以调整。

1. 冠词的减译

英语中有冠词，而汉语中没有这类词语。冠词是英语名词之前常使用的一种虚词，其本身没有独立的实际意义，通常有定冠词（the）和不定冠词（a/an）之分。翻译时，冠词省略不译的情况很多。例如：

(1) Secondary treatment can remove between 85 and 95 percent of *the* BOD and TSS in raw sanitary sewage.

译文：二级处理可去除生活污水中85%~95%的BOD和TSS。（减译定冠词the）

(2) If they are discharged into *a* lake, algal blooms and accelerated lake aging or eutrophication may be the result.

译文：如果它们被排放到湖中，可能会导致藻类大量繁殖，加速湖泊老化或富营养化。（减译了不定冠词 a）

(3) Another kind of rectifier consists of *a* large pear-shaped glass bulb from which all the air has been removed.

译文：另一种整流器由一个大的梨形玻璃灯泡构成，泡内的空气已全部抽出。

此外，有些情况下，冠词也需要译出。如不定冠词表示数量、表示每一、表示同一、固定搭配，定冠词表示特指、重复等情况下，不能减译。如：

(4) Such models have *a number of* special features, of which perhaps the most important are complete dynamic freedom and the absence of wind tunnel support and wall.

译文：这些模型具有一系列特点，其中最重要的特点是具有完全的动力自由度，没有风洞支架干扰和洞壁干扰。

2. 介词的减译

大量使用介词是英语的特点之一，它和冠词的使用频率一样高。虽然介词不能单独作为句子成分存在，但它可以表示名词或代词等与句中其他词的关系。汉语里面介词较少，而且用法也没有英语介词那么复杂。在翻译时，要么将介词转译为汉语里面的其他成分，要么省略不译。如：

(1) The difference *between* the two machines consists in power.

译文：这两台机器的区别在于功率不同。（减译介词between）

(2) The ionization in the upper atmosphere is caused by ultraviolet rays *from* the sun.

译文：上层大气中的电离作用是太阳的紫外线引起的。（减译介词from）

但是，需要注意的一点是，有些时候，介词非译出来不可，如：

(3) Electric current flows through a wire like tap water does *through* a pipe.

译文：电流通过导线流动就像自来水在管子内流动一样。（译出介词through）

3. 连词的减译

英语属于形合语言，其句式的构成需要连词的连接作用，常用于表示前后词语的逻辑关系和语法关系。相对而言，汉语中连词用得不多。一般来说，英语中的连词可以分为并列连词和从属连词两种，后者包括时间连词、地点连词、原因连词、条件连词、程度与结果连词，它们在句子中只是起到连接作用而不具有句子成分的功能，用来表达句子的逻辑关系。汉语则将时间、逻辑关系暗含在句子中，不需要使用连词。如：

(1) Check the circuit before you begin the experiment.

译文：检查好线路后再开始做实验。

(2) Like charges repel each other *while* opposite charges attract.

译文：同性电荷相斥，异性电荷相吸。（减译并列连词 while）

(3) The volume of a given weight of gas varies directly the absolute temperature *provided* the pressure does not change.

译文：压力不变，一定质量的气体的体积与绝对温度成正比。（减译条件连词 provided）

4. 引导词 there 的减译

英语中，以 there（也做副词）这个引导词引导的句子随处可见，可以与系动词以及其他部分不及物动词，如 seem，appear，exist，happen，stand，remain 等连用，构成与汉语不同的一种"某地存在/有……"句式。在这种句式中"there"已经失去了原有的意义，翻译的时候也大可不必强行翻译出来。

(1) *There* are three basic types or modes of land treatment: slow rate, rapid infiltration, and overland flow.

译文：土地处理有三种基本类型或方式：慢渗、快渗和地表漫流。（减译引导词 there）

(2) *There* exist neither perfect insulators nor perfect conductors.

译文：既没有理想的绝缘体，也没有理想的导体。（减译引导词 there）

5. 引导词 it 的减译

"it"除了做人称代词、指示代词和无人称代词之外，还可以用作引导词。此外，"it"不仅可以做形式主语，还可以起强调作用，改变句子的结构，还可以做形式宾语。例如：

(1) *It* is most often accomplished using a granular-media filter, much like the filters used to purify drinking water.

译文：颗粒介质过滤器是最常用的，像用于净化饮用水的过滤器。（减译形式主语 it）

(2) Scientists have proved *it* to be true that the boot get from coal and oil comes originally from the sun.

译文：科学家已运实，我们从煤和石油中得到的热都来源于太阳。（减译形式宾语 it）

二、意义需要进行的减译

和表示语法结构的虚词不同的是，实词是指具有实际独立意义并能在句子中充当句子成分的词，它包括名词、动词、代词、形容词、副词和数词等。具有意义的实词组成了英语句式的主干成分，一般需要翻译出来。然而，在保证原文和译文的意义不会流失，不影响理解的条件下，也可对实词进行减译。如：

1. 名词的减译

由于英语句法结构而不得不添加的名词或某种特殊结构中的名词，考虑到汉语表达时，不必翻译出来，而句子的意思也不会受损。如：

(1) The *problem* of alternative fuels of vehicle is one problem we shall approach.

译文：车辆的代用燃料是我们将要着手处理的一个问题。（减译名词 problem）

(2) The hardest *part* of any big project is to begin.

译文：任何一个大的项目最艰难的是起步阶段。（减译名词 part）

2. 动词的减译

一般来说，英语的句式比较固定，由"主语+谓语+宾语+……"构成，而汉语则可以使用意合的词语（包括其他词性的词，如名词、形容词、介词短语等）替代谓语动词。所以，将某些动词（包括行为动词和系动词，如 be / become /get 等）省略不译也是常见的。如：

(1) This dioxide *produces* about ten times more radiant power than that one.

译文：这只二极管的耗散功率比那只大9倍左右。（减译动词 produce）

(2) When the pressure gets low, the boiling point *becomes* low.

译文：气压低，沸点就低。（减译动词 become）

3. 代词的减译

英语中的代词极多，用来避免重复名词，包括人称代词、物主代词、反身代词、关系代词、指示代词等，而且有一些还有词形变化。翻译时，有的代词就可省略不译，例如：

(1) The denitrifying bacteria Pseudomonas and other groups use the carbon from the methanol and the oxygen from the nitrates in *their* metabolic processes.

译文：反硝化假单胞菌和其他菌群利用甲醇中的碳和硝酸盐中的氧进行代谢。（减译人称代词 their）

(2) But his findings gave some support to the idea *that* fusion may be possible without extreme heat.

译文：可是，他的发现支撑了这一想法：聚变可以在没有极高的温度下产生。（减译关系代词 that）

(3) The gas distributes *itself* uniformly throughout a container.

译文：气体均匀分布在整个容器中。（减译反身代词 itself）

4. 形容词减译

形容词一般是用来修饰说明名词的特性，但对于一些名词前显而易见的形容词可以省去不译，确保译文简洁通畅。如：

(1) This treatment did not produce any *harmful* side effect.

译文：这种治疗方法并没有产生任何副作用。（减译了形容词 harmful，因为副作用是有害的，不需要翻译为"有害的副作用"）

(2) It is quite hard to watch the *sparkling* stars clearly with our naked eyes.

译文：很难用我们的肉眼看清星星是怎样的。（减译了形容词"sparkling"，因为一般来说，星星是闪烁的）

三、修辞逻辑需要引起的减译

有些时候，英汉语言的差异非常微妙，多一个词或少一个词都显得格外别扭。修辞逻辑不仅会引起增译，也会引起必要的减译，省去拖沓的词或短语，让译文简明晓畅。如：

(1) Poison to a snake is merely a luxury; it enables it to get its food with very little *effort*, on more effort than one bite.

译文：毒液让毒蛇如虎添翼；有了毒液，捕食不费吹灰之力，也就是咬一口。（减译了 effort，译出"one bite"的含义可表示"effort"的含义）

(2) Archaeologists say the find may solve some of the mysteries *surrounding* the Inca civilization.

译文：考古学家称此次发现也许会解开印加文明之谜。（减译 surrounding，不然语言就不自然）

(3) A blood test showed that the level of *a fatty substance called* cholesterol in his blood was very high.

译文：验血表明他血液中的胆固醇含量相当高。（减译同义词短语 a fatty substance called）

Exercises 2

Translate the following paragraph into Chinese.

(1) The rotating drum is partially subemerged in the secondary effluent, which must flow into the drum and then outward through the microscreen. As the drum rotates, captured solids are carried to the top, where a high-velocity water spray flushes them into a hopper mounted on the hollow axle of the drum.

(2) Tertiary treatment of sewage can remove more than 99 percent of the pollutants from raw sewage and can produce an effluent of almost drinking water quality. But the cost of tertiary treatment, for operation and maintenance as well as for construction, is very high, sometimes doubling the cost of secondary treatment. The benefit-to-cost ratio is not always big enough to justify the additional expense. Nevertheless, application of some form of tertiary treatment is not uncommon. Some of the more common tertiary processes are discussed in the following sections.

Part 3　Writing

学术论文的撰写（四）——学术论文初稿写作要领

论文初稿的写作是属于定稿前的最初的草稿。初稿应包含的是所提出的主要观点以及对文字的初步推敲。在写初稿时，一般有四个步骤。

1. 将整个写作任务分解成若干个可操作的部分。

按照之前完成的论文提纲，制订出写作的详细提纲。详细的提纲就是比照初始的提纲把每一个写作任务具体化。这样，学术论文写作就从一项极费力气的工作变成可操作的，循序渐进的过程。

2. 一次只写一个部分。

从详细的提纲中选择某一部分开始写作，而且要争取一气呵成。千万不要在初稿时就反复推敲字句、修辞和文体，要集中精力在可能的时间内完成整个部分的写作。这时的写作线条可以粗一些，可以忽略某些词语，也可以不使用引语，在修改时再予补充。

3. 先写比较容易的部分。

这是接受度最广的一种写作方法。先完成较为容易的部分，然后，再对较难的部分进一步细分。在写作较容易的段落时，我们的工作方法和写作技巧都会得到更多的练习，而这些练习的成果会在进行较难部分写作时给予我们帮助和启迪。

4. 避免滥用文献资料。

这是一种科技英语写作特有的弊病。在准备写作时，收集大量的资料是必须的，也是必要的。但如果我们没有仔细地梳理并消化这些文献资料，就会带来一些副作用，最明显的副作用就是滥用文献资料。以下是几种滥用文献资料的常见情况。

（1）无须引证时，却详细地进行了引证；
（2）使用并不重要的背景材料；
（3）使用并不重要的引证，而且对资料没有进行正确处理；
（4）写作的内容偏离了主题；
（5）没有使用参考文献来验证某一观点，而是发现了另一新的内容；
（6）不知如何引用，只是感到资料的内容太为广泛，难以在某项具体的写作中使用，最后干脆放弃使用这些资料。

使用一定量的参考资料在科技写作中是必需的。但是，若边写作边阅读有关资料，会被大量的资料搅乱思路，写作也会很容易偏离原有的轨道和方向。要避免这种情况的产生就要消化最基本的资料，然后在写作时除了必须要证实某一论点或扩展某一内容外，不再另外查阅资料。

在写完初稿后，再对照提纲进行检查，看是否包括了所有的要点，是否已经扩展了所有的论点，是否还需要图表或其他细节，是否有结论等。

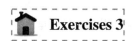

Exercises 3

Based on the academic article that you have taken the outline, please try to compile a first draft by yourself. Then compare it with the original article and find valuable differences between them.

Expanding Reading

What's Really in Your Water?

When water is safe, there is nothing better to drink. It's good for teeth, skin, weight control and even the ability to think straight. But drinking water contaminated with pathogenic bacteria, heavy metals or other harmful substances can cause diarrhea, brain damage, infertility and cancer.

Bottled water is no guarantee of water safety. Not only is it not always safe; bottled water is thousands of times more expensive than tap water, and the plastic packaging and transport carries heavy environmental costs.

Despite the need for safe water and the myriad pollutants that can contaminate it, there is no widely accessible method for everyone to quickly, cheaply, and accurately test their water's safety.

Nearly 50 years ago the same was true with pregnancy testing. Early advertisements for at-home pregnancy tests argued that women have "the right to know" if they were pregnant "with the least possible fuss and bother in the least possible time." This should also apply to drinking water quality.

Because the current water safety tests are still cumbersome to use for most people, the goal is to make them as easy to use as a pregnancy test so that they can be easily deployed in homes, day care centers and schools.

As professors at Northwestern University who research how to ensure water security for all, we are working on water testing by developing a new kind of test for drinking water that is rapid, cheap and accurate; and on quantifying water insecurity globally.

The intention is to implement this new water test into a format that nonscientists can easily use; one that is affordable and gives results within an hour for those who need them most. The technology is far from ready to sell; there is still much work to do to ensure that the lead tests are maximally user friendly.

Despite the need for safe water and the myriad pollutants that can contaminate it, there is no widely accessible method for everyone to quickly, cheaply, and accurately test their water's safety.

These tests are different because they harness the power of naturally occurring sensors from biology. Using tools from the nascent field of synthetic biology, the sensors can be programmed to change color when a target chemical is present in water.

These tests can be used to quickly detect lead, copper, arsenic and fluoride in water, with more analytes on the horizon. Northwestern and a spinoff company one of us co-founded are partnering to detect COVID-19 in sewage.

Global water quality has been jeopardized by a range of catastrophes both invisible and unmissable: failing infrastructure, massive fires, sewage overflows, agricultural runoff and forever chemicals sprayed in unexpected places.

Given the very real risks of drinking contaminated water, accurate knowledge about water quality is imperative for both action and advocacy. This necessitates cheap, fast and reliable methods of at-home testing that everyone can use.

(https://www.scientificamerican.com/article/whats-really-in-your-water/)

Unit 4 Solid Wastes and Soil Pollution

Lesson 1   Solid Wastes and the Management of Solid Wastes

Part 1   Reading

1 Solid Wastes

Solid waste is the unwanted or useless solid materials generated from combined residential industrial and commercial activities in a given area. Knowledge of the types of solid wastes, along with data on the composition and rates of generation, is basic to the design and operation of the functional elements associated with the management of solid wastes[①]. The materials that are collected under the term solid waste include many different substances from a multitude of sources. The sources of solid wastes are dependent on the socioeconomic and technological levels of a society.

● A small rural community may have known types of solid wastes from known sources (i.e. the wastes are more homogenous). Wastes from industrial and mining areas are also more homogenous.

● Urban communities (metropolitan cities) have many sources (the wastes are more heterogeneous).

1.1   Definition of Some Types of Solid Wastes

(1) Refuse: It is a general name given to all wastes except liquid waste. It includes all putrescible (decompose rapidly by bacteria) and non-putrescible (nondecomposable) wastes.

(2) Garbage: Putrescible wastes resulting from the growing, handling, processing, cooking and consumption of food, e. g. Vegetables, fruits, bones, bread, etc.

① High quantities of garbage are generated during the summer months, when vegetable wastes are more abundant.

② The increased use of processed and packaged foods has reduced garbage production and increased the combustible rubbish. It requires careful handling with frequent removal and adequate.

(3) Rubbish: It represents all non-putrescible wastes except ash. There are two categories of rubbish:

① Combustible: organic in nature and includes items such as paper, cardboard, wood yard clippings, bedding, plastics, etc.

② Non-combustible: are inorganic materials, which include metals, glass, ceramics, and other minerals.

(4) Ashes: It is an incombustible material that remains after a fuel or solid waste has been burnt.

(5) Infectious wastes: They are wastes that contain or carry pathogenic organisms in part or in whole such as wastes from hospitals and biological laboratories soiled with blood or bodily fluids.

(6) Special wastes: They are wastes from residential and commercial sources that includes.

① Bulky items (large worn out or broken household, commercial, and industrial items like furniture, lamps, bookcases, filing cabinets, etc.).

② Consumer electronics (includes worn-out, broken, and other no-longer wanted items such as radios, stereos, TV sets).

③ White goods (large worn-out a broken household, commercial, and industrial appliances such as stoves, refrigerators, dishwashers, clothes washers and dryers).

(7) Organic waste: Food waste, paper, cardboard, plastics, textiles, rubber, leather, wood, yard wastes.

(8) Inorganic waste: glass, crockery (cups, plates, etc.) tin cans, aluminum, and other metals.

(9) Dead bodies: Dead animals like dogs, cows, donkey, etc.

1.2 Composition of Solid Wastes and Their Determination

Composition is the term used to describe the individual components that make up the solid waste stream and their relative distribution, usually by percent by weight. Information on the composition of solid waste is important in evaluating equipment needs, systems and management programs and plans.

The types (components) of municipal solid waste may be different from country to country by season, economic condition, developmental level, etc.

1.2.1 Physical composition of solid wastes

The individual component study involves achieving the present composition of solid waste by volume and by weight. Volume measurements although difficult to measure are essential to disposal methods, e.g. to calculate incinerator sizes and land fill areas and to limit hauling capacity of refuse tracks, etc[②].

1.2.1.1 Moisture Content

The moisture content of solid waste is usually expressed in one of two ways:

- Wet-weight method of measurement: the moisture in a sample is expressed as a percentage of the wet weight of material.

- Dry-weight method of the measurement, it is expressed as a percentage of the dry weight of the material.

The moisture content of municipal solid waste varies depending on:
- Composition of the waste.
- The season of the year.
- Humidity.
- Weather condition esp. rain.

1.2.1.2 Density

Under physical composition of solid wastes density is one of the important parameters. Density is defined as the weight of the material per unit volume. The interest in knowing density of solid waste is to assess the total mass and volume of waste that must be managed. The densities of solid waste vary markedly with:
- Geographic location.
- Season of the year.
- Length of time in storage.

1.2.2 Chemical composition

Information on the chemical composition of solid wastes is important in evaluating alternative processing and recovery options. For example, the feasibility of combustion depends on the chemical composition of solid waste. If solid wastes are to be used as fuel, the four most important properties to be known are:

(1) Proximate analysis: measuring the total concentration of carbohydrate, protein, and lipid contents from solid wastes.

(2) Fusing point of ash: is defined as the temperature at which the ash resulting from the burning of waste will form a solid (clinker) by fusion and agglomeration. Typical fusion temperature for the formation of clinker from solid waste ranges from 1,100 to 1,200℃.

(3) Ultimate analysis: the ultimate analysis of a waste component typically involves the determination of the percent of C (carbon), H (hydrogen), O (oxygen), N (nitrogen), S (sulfur) and ash. The results of the ultimate analysis are used to characterize the chemical composition of the organic matter in municipal solid waste.

(4) Energy content: the energy content of the organic components in municipal solid waste should be determined.

1.2.3 Biological properties of MSW

The most important biological characteristic of the organic fraction of municipal solid waste is that almost all of the organic components can be converted biologically to gases and relatively inert organic and inorganic solids.

In planning for future waste management systems, it will be important to consider the changes it may occur in the composition of solid waste with time. Four waste components that have an important influence on the composition of the wastes collected are food waste, paper and cardboard, yard waste, and plastics.

2 Solid Waste Management

Management of solid waste reduces or eliminates adverse impacts on the environment and human health and supports economic development and improved quality of life. A number of processes are involved in effectively managing waste for a municipality. These include monitoring, collection, transport, processing, recycling and disposal.

2.1 Reduce, Reuse, Recycle

Methods of waste reduction, waste reuse and recycling are the preferred options when managing waste. There are many environmental benefits that can be derived from the use of the methods. They reduce or prevent greenhouse gas emissions, reduce the release of pollutant conserve resources, save energy and reduce the demand for waste treatment technology and landfill space. Therefore it is advisable that these methods be adopted and incorporated as part of the waste management plan.

Waste reduction and reuse of products are both methods of waste prevention. They eliminate the production of waste at the source of usual generation and reduce the demands for large scale treatment and disposal facilities. Methods of waste reduction include manufacturing products with less packaging, encouraging customers to bring their own reusable bags for packaging, encouraging the public to choose reusable products such as cloth napkins and reusable plastic and glass containers, backyard composting and sharing and donating any unwanted items rather than discarding them. All of the methods of waste prevention mentioned require public participation. In order to get the public onboard, training and educational programmes need to be undertaken to educate the public about their role in the process. Also the government may need to regulate the types and amount of packaging used by manufacturers and make the reuse of shopping bags.

Recycling refers to the removal of items from the waste stream to be used as raw materials in manufacture of new products. Thus from this definition recycling occurs in three phases: first the waste is sorted and recyclables collected, the recyclables are used to create raw materials. These raw materials are then used in the production of new products.

2.2 Treatment & Disposal

Waste treatment techniques seek to transform the waste into a form that is more manageable, reduce the volume or reduce the toxicity of the waste thus making the waste easier to dispose of. Treatment methods are selected based on the composition, quantity, and form of the waste material. Some waste treatment methods being used today include subjecting the waste to extremely high temperatures, dumping on land or land filling and use of biological processes to treat the waste. It should be noted that treatment and disposal options are chosen as a last resort to the previously mentioned management strategies reducing, reusing and recycling of waste.

2.2.1 Thermal treatment

This refers to processes that involve the use of heat to treat waste. Listed below are descriptions of some commonly utilized thermal treatment processes.

2.2.1.1 Incineration

Incineration is the most common thermal treatment process. This is the combustion of waste in the presence of oxygen. After incineration, the wastes are converted to carbon dioxide, water vapor and ash. This method may be used as a means of recovering energy to be used in heating or the supply of electricity. In addition to supplying energy incineration technologies have the advantage of reducing the volume of the waste, rendering it harmless, reducing transportation costs and reducing the production of the greenhouse gas methane.

2.2.1.2 Pyrolysis and gasification

Pyrolysis and gasification are similar processes they both decompose organic waste by exposing it to high temperatures and low amounts of oxygen. Gasification uses a low oxygen environment while pyrolysis allows no oxygen. These techniques use heat and an oxygen starved environment to convert biomass into other forms. A mixture of combustible and non-combustible gases as well as pyroligenous liquid is produced by these processes. All of these products have a high heat value and can be utilized

Gasification is advantageous since it allows for the incineration of waste with energy recovery and without the air pollution that is characteristic of other incineration methods.

2.2.2 Sanitary Landfills

Sanitary Landfills are designed to greatly reduce or eliminate the risks that waste disposal may pose to the public health and environmental quality. They are usually placed in areas where land features act as natural buffers between the landfill and the environment. For example the area may be comprised of clay soil which is fairly impermeable due to its tightly packed particles, or the area may be characterized by a low water table and an absence of surface water bodies thus preventing the threat of water contamination.

In addition to the strategic placement of the landfill other protective measures are incorporated into its design. The bottom and sides of landfills are lined with layers of clay or plastic to keep the liquid waste, known as leachate, from escaping into the soil. The leachate is collected and pumped to the surface for treatment. Boreholes or monitoring wells are dug in the vicinity of the landfill to monitor groundwater quality.

A landfill is divided into a series of individual cells and only a few cells of the site are filled with trash at any one time. This minimizes exposure to wind and rain. The daily waste is spread and compacted to reduce the volume, a cover is then applied to reduce odors and keep out pest. When the landfill has reached its capacity, it is capped with an impermeable seal which is typically composed of clay soil.

Some sanitary landfills are used to recover energy. The natural anaerobic decomposition of the waste in the landfill produces landfill gases which include carbon dioxide, methane and traces of other gases. Methane can be used as an energy source to produce heat or electricity. Thus some landfills are fitted with landfill gas collection (LFG) systems to capitalize on the methane being produced. The process of generating gas is very slow, for the energy recovery system to be successful there needs to be large volumes of wastes.

These landfills present the least environmental and health risk and the records kept can be a good source of information for future use in waste management, however, the cost of establishing these sanitary landfills are high when compared to the other land disposal methods.

2.2.3 Biological Waste Treatment

2.2.3.1 Composting

Composting is the controlled aerobic decomposition of organic matter by the action of microorganisms and small invertebrates. There are a number of composting techniques being used today. These include: in vessel composting, windrow composting, vermicomposting and static pile composting. The process is controlled by making the environmental conditions optimum for the waste decomposers to thrive. The rate of compost formation is controlled by the composition and constituents of the materials i.e. their Carbon/Nitrogen (C/N) ratio, the temperature, the moisture content and the amount of air.

C/N ratio is very important for the process to be efficient. The microorganisms require carbon as an energy source and nitrogen for the synthesis of some proteins. If the correct C/N ratio is not achieved, then application of the compost with either a high or low C/N ratio can have adverse effects on both the soil and the plants. A high C/N ratio can be corrected by dehydrated mud and a low ratio corrected by adding cellulose.

Moisture content greatly influences the composting process. The microbes need the moisture perform their metabolic functions. If the waste becomes too dry the composting is not favored. If however there is too much moisture then it is possible that it may displace the air in the compost heap depriving the organisms of oxygen and drowning them.

A high temperature is desirable for the elimination of pathogenic organisms. However, if temperatures are too high, above $75°C$, then the organisms necessary to complete the composting process are destroyed. Optimum temperatures for the process are in the range of $50\text{-}60°C$ with the ideal being $60°C$.

Aeration is a very important and the quantity of air needs to be properly controlled when composting. If there is insufficient oxygen, the aerobes will begin to die and will be replaced by anaerobes. The anaerobes are undesirable since they will slow the process, produce odors and also produce the highly flammable methane gas. Air can be incorporated by churning the compost.

2.2.3.2 Anaerobic Digestion

Anaerobic digestion like composting uses biological processes to decompose organic waste. However, where composting can use a variety of microbes and must have air, anaerobic digestion uses bacteria and an oxygen free environment to decompose the waste. Aerobic respiration, typical of composting, results in the formation of carbon dioxide and water. While the anaerobic results in the formation of carbon dioxide and methane. In addition to generating the humus which is used as a soil enhancer, anaerobic digestion is also used as a method of producing biogas which can be used to generate electricity.

🏠 Words and Expressions

homogenous	[hə'mɑdʒənəs]	*adj.* 同种类的，同性质的，由相同成分（或部分）组成的
heterogeneous	[ˌhetərə'dʒɪːnɪəs]	*adj.* 由很多种类组成的，各种各样的
putrescible	[pjʊ'trɛsəbl]	*adj.* 易腐烂的
fluids	['fluədz]	*n.* 液体，流体，液
clipping	[klɪpɪŋ]	*n.* 剪下物
ceramics	[sə'ræmɪks]	*n.* 陶瓷制品，陶瓷器，制陶艺术
bulky	['bʌlkɪ]	*adj.* 庞大的，笨重的
filing cabinet		*n.* 文件柜，档案柜
leather	['leðər]	*n.* 皮革
crockery	['krɑːkərɪ]	*n.* 陶器，瓦器，碟、盘、杯、碗
representativeness	[ˌreprə'zentətɪvnəs]	*n.* 代表性
randomization	['rændəmɪ'zeʃən]	*n.* 随机化，不规则分布，随机选择
hauling capacity		*n.* 牵引能力
moisture	['mɔɪstʃər]	*n.* 潮气，湿润，水分
proximate analysis		*n.* 近似分析
fusing point		*n.* 熔点
clinker	['klɪŋkər]	*n.* 煤渣，炉渣
agglomeration	[əˌglɑːmə'reɪʃn]	*n.* （杂乱聚集的）团、块、堆、聚集
ultimate analysis		*n.* 元素分析，最终分析
napkin	['næpkɪn]	*n.* 餐巾，餐巾纸
pyrolysis	[ˌsedɪmen'teɪʃn]	*n.* 高温裂解，热解，高温分解
placement	['pleɪsmənt]	*n.* （对物件的）安置，放置
incorporate	[ɪn'kɔːrpəreɪt]	*v.* 将……包括在内，包含，吸收，

			使并入
boreholes	[ˈbɔːˌhəʊlz]	n.	钻孔，井眼
in-vessel composting		n.	密封罐堆肥
windrow composting		n.	料堆堆肥
vermicomposting		n.	蚯蚓堆肥，蚯蚓堆制处理
humidity	[hjuːˈmɪdətɪ]	n.	（空气的）湿度，湿热，高温潮湿
dehydrate	[dɪˈhaɪdreɪt]	v.	脱水
cellulose	[ˈseljuloʊs]	n.	纤维素
heap	[hɪːp]	n.	（凌乱的）一堆，许多，大量；
		v.	堆积（东西），堆置
churning	[ˈtʃɜːrnɪŋ]	adj.	（水）剧烈翻滚的，湍急的；
		v.	剧烈搅动，（使）猛烈翻腾
humus	[ˈhjuːməs]	n.	腐殖质

Notes

① Knowledge of the sources and types of solid wastes, along with data on the composition and rates of generation, is basic to the design and operation of the functional elements associated ith the management of solid waste.

参考译文：掌握固体废弃物来源、种类、成分以及产生量的相关数据是固体废弃物管理要素设计和运行的基础。

② Volume measurements although difficult, they are essential to disposal methods, e.g. to calculate incinerator sizes and land fill areas and to limit hauling capacity of refuse tracks, etc.

参考译文：体积测量虽然很难，但却是选择处置方法的基础，比如，它可以用来计算焚烧炉的尺寸和填埋场的面积，限定垃圾车的牵引能力。

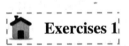

Exercises 1

1. According to the reading material, chose the best answer(s) from the options.

(1) According to the passage, food waste is regard as _____ .
   A. rubbish   B. garbage   C. special wastes   D. none of them.

(2) The types (components) of municipal solid waste may be different from country to country by _____ , _____ and _____ .
   A. season   B. economic condition
   C. developmental level   D. climatic conditions

(3) Under physical composition of solid wastes, the important parameters are _____ and

_____.

A. putrescible   B. humidity   C. density   D. moisture

(4) A number of processes are involved in effectively managing waste for a municipality. These include _____ , _____ , _____ , processing, and disposal.

A. monitoring   B. collection   C. transport   D. recycling

(5) Thermal Treatment refers to processes that involve the use of heat to treat waste, including _____ , _____ and gasification

A. incineration   B. landfill   C. composting   D. pyrolysis

(6) _____ present the least environmental and health risk.

A. Incineration   B. Landfill   C. Composting   D. Pyrolysis

(7) The rate of compost formation is controlled by the composition and constituents of the materials, such as _____ , _____ , _____ and _____ .

A. C/N ratio   B. the temperature
C. the moisture content   D. the amount of air.

2. Propose how to process the solid waste by sanitary landfill.

3. Propose how to process the solid waste by biological waste treatment.

Part 2 Translation

英语翻译中的数量词翻译

数量词作为信息的重要载体，在科技类文章中有着举足轻重的作用，准确无误地翻译数量词是保证译文质量的重要前提。科技英语中数词的表达方式有时与汉语存在较大差别。英语中的数词可以用作主语、表语、宾语和定语等，翻译时可能会遇到困难，甚至出现误解和错译。

英语中表示增加的动词有 increase/rise/grow/gain 等；表示减少的动词有 decrease/reduce/lower 等，当这些词后面跟介词 to 时，翻译时采用直译的方法译成"增加到……"或"减少到……"，当这些词后面跟 by 时，表示净增减，可直译为"增加了……"或"减少了……"。

用"as...as..."结构也能表示数量的增减。例如：

(1) The outer portion of the wheel may travel as fast as 600 miles per hour.

译文：轮子外缘的运动速度可能高达每小时 600 英里（1 mile=1.61 km）。

有时，也用"by+名词+形容词比较级"表示数量的增减。例如：

(2) The wire is by 10 inches shorter than the one you used.

译文：这根导线比你用的那根短 10 英寸（1 in=2.54 cm）。

另外，还有一些表示增加或减少的短语及译法：

as many/much as　与……一样多

five times as many/much as　是……的5倍

to jump 50 percent above the previous year　比去年增长50%

to go up by…; to shoot up by…; to rise by (to)…; to be raised by 增加了……

to go down by…; to fall by（to）…; to reduce by… 下降，减少了……

a reduction/fall of 70 percent 下降（减少）70%

reduce by 1 percent; to decrease by one-hundredth 下降1%

to decrease/reduce by one-fifth; to reduce to four-fifths; to be reduced to four fifths 下降20%（或1/5）

to decrease/reduce by one quarter; to reduce to three-fourths 下降25%（或1/4）

一、数字的翻译

英语和汉语的数字表达既有相同的部分，也有不同的部分，如两者的进位都是从右到左，即从个位开始，但在英语中每三位为进位制的一组，并用逗号隔开，从右开始，第一个逗号前为"千"（thousand），第二个逗号前为"百万"（million），以此类推。一般来说，对于具体的数字，可直接翻译出来。例如：

(1) Typical fusion temperature for the formation of clinker from solid waste ranges from 1,100 to 1,200℃.

译文：通常焚烧固体废物的温度范围为1100~1200 ℃。

例句中的数字直接译出即可。这类可以直接译出的数字一般不大，包括表温度、年代、数量、高度等，但对于较大的数字，翻译时需要按照英汉语言表式进行转换，如数字657000用英语表示应为 six hundred and fifty-seven thousand，用汉语表示则应为"65万7千或六十五万七千"。例如：

(2) Half-lives of different radioactive elements vary from as much as 900 million years for one form of uranium, to a small fraction of a second for one form of polonium.

译文：不同的放射元素，其半衰期也不同，有一种铀的半衰期长达9亿年，但有一种钋的半衰期却短到几分之一秒。

例句中将900 million转换成汉语中的9亿，容易理解。值得注意的是，科技英语中还大量使用不定数量词，用来指若干、许多、大量、不少、成千上万等概念的词，例如：

(3) The number of known hydrocarbons runs into tens of thousands

译文：已知的碳氢化合物多达几万种。

例句中使用了 ten、 thousand 数次，其后加复数后缀s，形成一般的约定俗成的译法如 hundreds of（几百、成百上千）、 thousands of（几千、成百上千）等，类似的词还有很多，如 number、lot、score、 decade、 million 等。

另外，还有表示"多达""不足""左右""到"等数量介词，如 above、 more than、over、below、less than、under、around、close to、nearly、some 等，翻译时，应结合原文仔

细推敲。

二、倍数的翻译

科技英语中，使用频率较高的一种数量增减就是倍数表达。英语中表示倍数增加的形式多种多样，而汉语中通常只有两种表达方式，一种是增加了多少倍，另一种是增加到原来的多少倍。翻译时，要特别注意，避免失误。

1. 倍数增加的译法

在翻译倍数的增加时，一定要注意所增加的倍数是否包含基数。如包含基数，通常可译为"增加到……倍""增加为……倍"等；如不包含基数表示净增，可译为"增加了……倍"。常用表达及译法有四种：

increase by n times "增加了 n-1 倍"或"增加到 n 倍"

increase n times "增加了 n-1 倍"或"增加到 n 倍"

increase by a factor of n "增加了 n-1 倍"或"增加到 n 倍"

increase to n times "增加了 n-1 倍"或"增加到 n 倍"

例如：

(1) The strength of the attraction increases by four times if the distance between the original charges is halved.

译文：如果原电荷之间的距离缩短一半，引力就会增加 3 倍。（或：就会增加到原来的 4 倍。）

(2) The production of this year is estimated to increase to 3 times compared with 2004.

译文：和 2004 年相比，今年的产量预计增长了 2 倍。（或：是 2004 年的 3 倍）。

另外，还有一些表示倍数增加的短语及译法，例如：

(to) double 双倍，加倍

(to) treble/ triple 三倍

(to) quadruple 四倍，翻两番

to be double that of last year 是去年的 2 倍

to be more than double the 2006 figure 比 2006 年增加 1 倍以上

to be a dozen times that of 2005 比 2005 年增加了 10 多倍

2. 倍数减少的译法

英语可以用 times 等表示倍数的词语来表示数量的减少，即"减少了 n 倍"或"成 n 倍地减少"。由于汉语不能说"减少了多少倍"，而习惯讲"减少了几分之几"，所以翻译时应该进行换算。此外，当英语中的倍数为小数时，则应该化为整数，如：shorten/ reduce 3.5 times，应该化为（3.5-1）/3.5=5/7，译为"缩短/减少了 5/7"；或者 1/3.5=2/7，译为"缩短减少到 2/7"。常用表达及译法有五种：

decrease by 3 times 减少了 2/3 （减至 1/3）

decrease by a factor of 3 减少了 2/3 （减至 1/3）

decrease 3 times 减少了 2/3 （减至 1/3）

3 times less than 减少了 2/3 （减至 1/3）

A is 3 times smaller than B　　A 是 B 的 1/3 或 A 比 B 小 2/3

例如：

(1) The weight of the electronic device has decreased by four times.

译文：这种电子器件的重量减轻了 3/4。

(2) The power output of the machine is twice less than its input.

译文：该机器的输出功率比输入功率小 1/2。

(3) The wire is two times thinner than that.

译文：这根导线比那根导线细 1/2。

3. 倍数比较的译法

倍数比较在科技英语中使用频率也较高，常用表达及译法有四种：

A is n times larger than B　　A 是 B 的 n 倍或 A 比 B 大 n-1 倍（净增 n-1 倍）

A is n times as large as B　　A 相当于 B 的 n 倍或 A 比 B 大 n-1 倍（净增 n-1 倍）

A is larger than B by n times　　A 是 B 的 n 倍（净增 n-1 倍）

A is n times B　　A 是 B 的 n 倍（净增 n-1 倍）

例如：

(1) The oxygen atom is 16 times heavier than the hydrogen atom.

译文：氧原子的重量是氢原子的 16 倍。

另外，还有一种特殊的表达法，即 be/$v.$+ as +$adj.$ again as 译作"是……的两倍"，again 前面加上 half，则表示"是……的一倍半"。

例如：

(2) This wire is as long again as that one.

译文：这根线是那根的 2 倍。（或：比那根长 1 倍。）

三、百分比的翻译

科技英语中表示百分比增减的方式很多。

1. 百分数+形容词比较级+than

该结构表示净增减概念，百分比可以直接译出来，例如：

The can sealing machine works fifty percent faster than that one.

译文：这台封罐机比那台的工作速度快 50%。

2. 百分数+形容词比较级+名词

该结构表示净增减的数量，百分数可直译出来，例如：

This new-type pump wasted 20 percent less energy supplied.

译文：这台新型水泵少损耗20%的能量。

(2) An increase in lubricant temperature results in 20 percent lower viscosity.

译文：由于润滑剂温度增高，黏度降低20%。

3. 表示增减意义的动词+ by +百分数

该结构表示净增减，百分数可直译出来，例如：

New booster can increase the pay load by 120%.

译文：新型助推器能使有效负荷增加120%。

4. 表示减少意义的动词+ to +百分数

该结构可以用来表示减少后剩余的数量，应译为"减少百分之……"，例如：

(1) By the year 2003, the world's annual oil output is expected to fall to 30%.

译文：到2003年，全世界年产油量预计将下降到30%。

此外，还有几种表示百分比的表达法，例如：

(2) This month we have produced 110 percent the number of transformers last.

译文：这个月的变压器产量是去年的110%。

例句中使用了"百分数+名词/代词+ of +名词"表达法，该结构表示的增减包括底数在内。

还有分数的表示法也值得一提，其一般的公式为：分子（用基数词）/分母（用序数词复数）。当分子小于或等于1时，分母用单数形式，如1/3（one third），但1/2则用one（或a）half表示，而不用 one second。当表示A是B的几分之几，或A比B小几分之几时，则用A is+倍数+ as large as B 或 A is+倍数+ smaller than B，如：A is one fourth as large as B；A is two-thirds smaller than B.

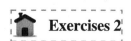

Translate the following paragraph into Chinese.

(1) Wet-weight method of measurement: the moisture in a sample is expressed as a percentage of the wet weight of material.

(2) A landfill is divided into a series of individual cells and only a few cells of the site are filled with trash at any one time.

(3) A high temperature is desirable for the elimination of pathogenic organisms. However, if temperatures are too high, above 75℃, then the organisms necessary to complete the composting process are destroyed. Optimum temperatures for the process are in the range of 50-60℃ with the ideal being 60℃.

Part 3　Writing

学术论文的撰写（五）——学术论文初稿的修改

初稿完成后就可以开展修改工作了。我们需要检查主题是否突出，内容是否准确，写作技巧是否运用得当，所引用的数据与图表是否正确，数字是否与引用的材料一致等。修改应从大到小进行，即从主题、结构等大框架到技巧词句等小细节。正文的修改也可以采用同样的思路。

一、主题修改

首先要初步检查主题是否有说服力，可以从检查文章的逻辑性、连贯性和流畅性开始。在初步修改中，主要做以下三个方面的工作：

1. 论文的主题是否清晰、一致、充分地表达

（1）前言部分是否清楚地说明了所要论述的主题。
（2）是否所有的支持段落都支持了主题。

2. 主题是否有足够的支撑

（1）是否至少有 3 个或 3 个以上从不同角度对论题进行支持的论点。
（2）对每一个论点是否都有明确的论证。
（3）对每一个论证是否都有足够的论据。

3. 结构是否严谨

（1）论文的结构是否有一个清晰的思路。
（2）前言是否能引起读者兴趣，结论是否完整，题目是否能准确表达主题。

二、结构修改

在对初稿的主题修改结束后，就要检查论文结构是否合理，此时，细节至关重要。如果发现了问题，就要再进行检查与核实。这时的主要修改工作要求如下：

1. 注意是否正确使用了"元话语"

检查文章结构，确保使用了恰当的"元话语"。"元话语"是指关于话语的话语，相当于对论文内容的预告，也可以理解为引导读者阅读的"路标"，可以起到提纲的作用，使文章逐步地按逻辑展现在读者眼前。在表示结束时，用 in conclusion 来表示对文章内容的总结；in addition 或 furthermore 可以表示论点之间的相互关系；first, second, third 可以表明讨论内容的顺序。

2. 是否需要添加小标题

为了保证论文的一致性和结构的严谨性，可以使用小标题和二级标题。小标题不仅会给读者提供有效的信息，还会给作者提供文章的结构，如 Introduction, Procedure, Results, Conclusion 等，都可作为英语科技写作中常见的小标题。许多人把小标题作为提纲的一部分来使用，有人把小标题作为进一步修改文章结构的依据。

三、句段和修辞修改

在完成对论文结构检查的基础上，对句子、段落和修辞有以下 3 个必做的工作：

1. 用词是否保持了一致性

（1）使用动词，注意时态上要保持一致。如：

① Joan punched down the risen yeast dough in the bowl. Then she dumps it on the floured worktable and kneaded it into a smooth, shiny ball.（句中动词的时态不一致）

② Joan punches down the risen yeast dough in the bowl. Then she dumps it on the floured worktable and kneads it into a smooth, shiny ball.（经过修改，动词的时态保持一致的句子）

（2）使用代词注意前后一定要保持一致，否则就会造成所述的观点矛盾。如：

① One of the fringe benefits of my job is that you can use a company credit card for gasoline.（前后不一致）

② One of the fringe benefits of my job is that I can use a company credit card .（前后一致）

2. 在选词上是否意义明确

为了保证所写的论文内容不产生歧义，要使用能明确表达意义的词汇，以便能在读者的脑海里产生一个清楚的概念。如：

(1) The man signed the document.（所用词汇的意义比较笼统）

(2) The biology professor hastily scribbled his name on the course withdrawal slip.（所用词汇意义比较明确）

为了能准确地选择用词来表明意义，应尽量使用确切名称，如人名、地名和事物名。在名词和动词前要使用描述性的词汇，更准确地限定所用名词或动词的性质和特征，有时还需要使用定语从句或状语从句来说明。

3. 句型是否多样化

高质量的写作就是在文章中要使用不同的句型结构，避免单调乏味。一般来说，主动语态结构简单，在文体上显得更有活力；而被动语态结构略复杂但也较严谨，突显客观性。因此在选用句型时，要明确表达目的。简单句结构简明，不容易出错且表达更为明确有力。而选用并列句和复合句会在连接词的使用上体现内容的逻辑关系，有利于表达多层次的内容。如：

(1) Solid waste is the unwanted or useless solid materials generated from combined residential

industrial and commercial activities in a given area. （所用句型为主动语态的简单句，简洁有力。在宾语后面使用了动名词复合结构进一步说明 materials 的性质来源,令表达更加充分、准确）

(2) It should be noted that treatment and disposal options are chosen as a last resort to the previously mentioned management strategies reducing, reusing and recycling of waste. （所用句型为被动语态的复合句。强调了处理技术和方案本身各有特点，对其的选择在于人们的管理策略，客观、实际，更具说服力。宾语从句的使用突显了"技术无好坏，决策者的策略非常重要"的主题表达，又进一步联系上文内容，在逻辑上有明显的递进关系，严谨又清晰）

最后，为了让句段表达更具多样性,还可以使用一些特殊的词汇或短语开头。如-ed、-ing、-ly 结构的动名词、动词不定式或介词短语。如:

(1) Generally, this leaves 30 mg/L or less of BOD and TSS in the secondary effluent.

(2) To understand the effects of water pollution and the technology applied in its control, it is useful to classify pollutant into various groups or categories.

(3) In planning for future waste management systems, it will be important to consider the changes it may occur in the composition of solid waste with time.

在修改初稿的过程中，可以看出写作是没有固定程式的，想要写出一篇好论文不仅需要大量阅读的积累，还需要去揣摩优秀作者的谋篇布局、表达技巧和行文方式。这不是一蹴而就的，但在循序渐进的练习中，我们最终会把英文论文写作变得更加可控和可操作。

Exercises 3

Based on the academic article that you have taken the outline, please try to compile a first draft by yourself. Then compare it with the original article and find valuable differences between them, do your modifications on the first draft.

Tech Waste Is a Danger to Us All

Industry and governments are ignoring the problem—as well as some simple solutions

Hardware in general, and smartphones in particular, have taken over our lives so quickly that few of us have had the chance to think about what happens to them when we no longer use them. The answer is that they become a huge environmental and health problem in the Global South's landfill sites.

We don't hear much about this problem because it is out of sight, out of mind. Electronic waste is currently 5 percent of all global waste, and it is set to increase exponentially as more of us own multiple smartphones, laptops and power banks—few of which are likely to be repaired or recycled at the end of their lives.

There is 100 times more gold in a metric ton of mobile phones than in the same amount of gold ore; up to 7 percent of the world's gold may currently be contained in e-waste.

If it is economically viable to mine gold ore, it can be just as profitable to mine iPhones, with the right technology and processes.

This would also future-proof manufacturers' supply chains since there are concerns about the long-term availability of raw materials they need like gold, platinum, cobalt, aluminum and tin.

Before manufacturers even get to the costly process of developing these recycling and extraction systems, there is an opportunity for them to simply ask consumers to trade in their old phones. They could then act as brokers for resale of these items to emerging markets that are full of aspirational consumers eager to own their first smartphone—even if it's an iPhone 6, not an 11.

As well as the financial benefit, they will be able to capitalize on the public relations opportunity and block cheaper smartphone manufacturers from establishing a foothold and building customer loyalty in emerging markets.

These are policies that could be as popular in Silicon Valley as in Accra. It is high time our governments and industries started to push for them.

Unit 4 Solid Wastes and Soil Pollution

Lesson 2 Hazardous Waste Management

Part 1 Reading

1 Introduction

The management of hazardous wastes has become a specialized discipline because of the complex nature of the problem and the solutions available to us. The mismanagement examples of hazardous wastes causing disastrous human and environmental consequences are numerous. Therefore, the current emphasis is on "cradle to grave" waste management[①]. The management process begins with an understanding of the definition and classification of the different wastes, and their harmful effects on human health, and ends with the application of risk management to control human health and environmental impacts of hazardous waste.

With the rapid pace of urbanization and industrialization, waste generation activities resulting from the provision of goods and services have increased at an alarming rate throughout the world. The modern standards of living require the usage of products made and refined from various natural resources. Almost all the manufacturing processes generate solid, liquid and/or gaseous emission as by-products. Some of these emissions are potentially harmful to human health and environment, and thus warrant special techniques of management. Hazardous waste management, therefore, deals with minimizing harmful effects on humans and environment by applying special techniques of handling, storage, transportation, treatment and disposal of hazardous emissions. Current practice also calls for pollution prevention, a term used to describe source reduction and recycling activities contributing to a reduction in the volume of waste generated at the source.

Contaminated sites are the result of poor management of hazardous waste in the past. Therefore, effective hazardous waste management should minimize short-term risks immediately after generation of the waste and the long-term risks associated with potentially contaminated sites. Consequently, the process of hazardous waste management starts with problem definition or identification and characterization of waste, application of pollution prevention activities to minimize waste needing handling, storage, transportation and pollution control involving appropriate treatment.

The motivation to manage hazardous wastes to minimize human health and environmental risks came from a public awareness that grew over many historical incidents of uncontrolled disposal of hazardous waste. The most publicized landmark incident is worth recounting as reminder.

2 Love Canal, United States

The Love Canal incident[2] in the United States is a classic example of contamination of a site with hazardous waste over a period of time. As part of a hydroelectric and industrial project, a businessman named W. T. Love originally excavated the Love Canal in Niagara Falls, New York. However, the project was terminated prior to completion. The Hooker Chemicals and Plastic Company subsequently used the blocked-off canal as a chemical disposal site from 1942 to 1953. An estimated twenty thousand tonnes of chemical waste, much of it containing dioxin and other highly toxic compounds, was dumped there.

In 1953, Hooker Chemical and Plastic Company sold the site to the local Board of Education for one dollar, with the condition that the company would be cleared of any liability for the site[3]. An elementary school was built on the site, and the remaining land was sold to developers for housing developments. With time, Love Canal's water, soil and air were contaminated with hazardous chemicals. The hazardous contents within the canal started migrating into the surrounding soil and water environment. Even though the contents of the canal were highly toxic, it took a while before health problems were noticed and residents complained.

The site eventually became the first U. S. federal environmental disaster area. The level of on-site contamination was found to be 250 to 5,000 times the level deemed safe for human exposure. In 1978, two hundred and forty families were evacuated from the area. Subsequently, over five hundred homes near the site were also evacuated. There were abnormally high rates of miscarriages, birth defects and cancer among the former residents.

In 1990, after over a decade of studies, remediation and relocation costing over 275 million dollars, some of the area was reopened to new occupancy. The now deemed habitable part of Love Canal was renamed Black Creek village.

3 Lessons Learned

The example highlighted above and the many more similar cases in North America and elsewhere have contributed a vast pool of practical and scientific knowledge about the short-term and long-term effects of hazardous waste on human health and environment. Scientists have realized the need to develop clear cause and effect relationships between chemical agents and human disease. The role of production, storage, transport, disposal and transformation of chemical compounds in the manifestation of human health and environmental effects has become a major topic for scientific research.

Without scientific data to absolve chemical agents, the response of the international community was to implement stringent regulations to control hazardous wastes[④]. Pollution prevention was touted as the solution to all ills associated with hazardous waste. Considering the monumental costs associated with rectifying past mistakes or mismanagement, pollution prevention at a reasonable cost seems to make perfect sense.

Because of the mistake made in the past, the current approach to hazardous waste management follows the "cradle to grave" concept of comprehensive waste management. Implementation of this concept requires the early identification of hazardous wastes, and actions contributing to formation of such wastes, as well as a step-by-step approach to minimize risks associated with the production, handling, and eventual disposal of such waste.

4 Definition and Classification of Hazardous Waste

As a first step, a classification system is necessary to identify waste type capable of causing human health and environmental impacts. Unfortunately, a universal classification system, acceptable to all countries, is currently not available. Nevertheless, there is general agreement on a definition of hazardous waste based on the following functional properties:

- Ignitablility.
- Corrosivity.
- Reactivity.
- Toxicity.

In addition to identifying wastes based on these properties, specific waste types that are included on hazardous waste lists of local regulatory bodies will be automatically classified as hazardous[⑤]. In some cases, exceptions to the classification are also listed. With standard tests established by regulatory bodies such as the Environmental Protection Agency (EPA) in the United States, the defining parameters can be described as follows:

- *Ignitability* is the property that could cause a fire during routine management. Examples of ignitable wastes include waste oils and used solvents.
- *Corrosivity*, as indicated by highly acidic and alkaline materials, is a basic property of a hazardous waste because such corrosive materials can react dangerously with other wastes or cause toxic contaminants to migrate from certain wastes. An example of a corrosive waste includes used pickle liquor from steel manufacturing.
- *Reactivity* of unstable wastes can pose an explosive problem at any stage of the waste management cycle. Examples of reactive wastes include water from TNT operations and used cyanide solvents.
- *Toxicity* of wastes that are likely to leach hazardous concentrations of particular toxic

constituents into ground water aquifers is an important property for classification. In simulated leaching actions that occur in landfills, if the concentration of the toxic constituent exceeds a regulatory limit, the waste is termed hazardous. Furthermore, any waste exceeding regulatory acute toxicity limits (in terms of LC_{50}[⑥] or LD_{50}[⑦] values) is classified as a hazardous waste. Examples include wastes containing high levels of volatile organic compounds (e.g. benzene) and heavy metals such as lead, cadmium and mercury.

A regulatory classification system using the above criteria has a number of limitations. Relative differences in hazards among wastes are not examined by such a system. According to this system, once a waste is classified as hazardous, then it would be accorded the same regulatory coverage regardless of the degree of hazard.

Words and Expressions

disastrous	[dɪˈzæstrəs]	*adj.* 极糟糕的，灾难性的，完全失败的
provision	[prəˈvɪʒn]	*n.* 提供，供给，给养，（为将来做的）准备，（尤指旅途中的）粮食；*v.* 为……提供所需物品（尤指食物）
motivation	[ˌmoʊtɪˈveɪʃn]	*n.* 动机，动力，诱因
excavate	[ˈekskəveɪt]	*v.* 发掘，挖出（古建筑或古物），挖掘，开凿，挖空（洞、隧道等）
terminate	[ˈtɜːrmɪneɪt]	*v.* （使）停止，结束，终止，到达终点站
completion	[kəmˈplɪːʃn]	*n.* 完成，结束，（房地产等的）完成交易，完成交割
dioxin	[daɪˈɑːksɪn]	*n.* 二噁英，二氧（杂）芑
evacuate	[ɪˈvækjueɪt]	*v.* （把人从危险的地方）疏散、转移、撤离，（从危险的地方）撤出，排泄（粪便）
miscarriage	[ˈmɪskærɪdʒ]	*n.* 流产
birth defects		*n.* 先天畸形；出生缺陷
occupancy	[ˈɑːkjəpənsi]	*n.* （房屋、土地等的）占用、使用、居住
habitable	[ˈhæbɪtəbl]	*adj.* 适合居住的
manifestation	[ˌmænɪfeˈsteɪʃn]	*n.* 显示，表明，表示，（幽灵的）显现，显灵

touted	['taʊtɪd]	v. 标榜，吹捧，吹嘘
rectifying	['rektɪfaɪɪŋ]	v. 矫正，纠正，改正
comprehensive	[ˌkɑːmprɪ'hensɪv]	adj. 全部的，（几乎）无所不包的，详尽的，综合性的
implementation	[ˌɪmplɪmen'teɪʃən]	n. 执行，实施，贯彻，生效
regulatory	['regjələtɔːrɪ]	adj.（对工商业）具有监管权的，监管的
pickle	['pɪkl]	v. 腌渍
cyanide	['saɪənaɪd]	n. 氰化物（剧毒化学品）；vt. 用氰化法处理
leach	[liːtʃ]	v. 过滤，滤，滤去；n. 过滤，过滤器
benzene	['benziːn]	n. 苯
cadmium	['kædmɪəm]	n. 镉

Notes

① cradle to grave waste managment 从摇篮到坟墓废物管理，即将整个危险废物的产生到最后的处置视为人的整个生命周期。

② Love Canal incident 拉夫运河事件。拉夫运河位于纽约州，1942 年，美国一家电化学公司购买了这条大约 1000 m 长的废弃运河，当作垃圾仓库来倾倒大量工业废弃物。1953 年，这条充满各种有毒废弃物的运河被公司填埋覆盖好后转赠给当地的教育机构。此后，纽约市政府在这片土地上陆续开发了房地产，盖起了大量的住宅和一所学校。从 1977 年开始，这里的居民不断发生各种怪病，孕妇流产、儿童夭折、婴儿畸形、癫痫、直肠出血等病症也频频发生。

③ In 1953, Hooker Chemical and Plastic Company sold the site to the local Board of Education for one dollar, with the condition that the company would be cleared of any liability for the site.

参考译文：1953 年，Hooker 化学和塑料公司以 1 美元的价格将这块地出售给当地的教育委员会，其条件是公司不再对这块地承担任何法律责任。

④ Without scientific data to absolve chemical agents, the response of the international community was to implement stringent regulations to control hazardous wastes.

参考译文：如果没有科学的数据能为化学试剂洗脱罪责，国际社会的反应就应该是实施严格的法规来控制有害废物。

⑤ In addition to identifying wastes based on these properties, specific waste types that are included on hazardous waste lists of local regulatory bodies will be automatically classified as hazardous.

参考译文：除了根据上述特征定义废物外，包括在当地法规机构所列出的有害废物清单里的具体废物将会自动划分到有害类。

⑥ LC_{50}（Lethal Concentration 50%）致死中浓度/半致死浓度/半数致死浓度，表示杀死50% 防治对象的药剂浓度。

⑦ LD_{50}（Lethal Dose 50%）毒理学中的半数致死量。

 **Exercises 1**

1. According to the reading material, chose the best answer(s) from the options.

(1) According to the passage, which of the followings were found among the former residents of contaminated site of Love Canal? _____

A. Neurological disorder   B. Miscarriages   C. Birth defects   D. Cancer

(2) Which of the followings is (are) NOT regarded as the property of hazardous wastes? _____

A. Biodegradable   B. Corrosivity   C. Toxicity   D. Ignitability

(3) The definition of hazardous wastes is universal all over the worlds. _____

A. The statement is TURE.   B. The statement is FALSE.

2. Use another expression to replace "cradle to grave".

3. Explain the LC_{50} in English.

Part 2　Translation

英语翻译中的词性转换

词性转换是常用而且高效的翻译技巧之一，它有助于消弭英汉差异所引起的语言的机械化转换。英语中有动名词、不定式、分词、冠词和关系代词等，而汉语中则不一定会有这些词，即便有，功能也未必相同。对于这些不对应甚至是没有的词，则要按照汉语表达特征进行适当的调整和转换，这样才会有妥帖、顺畅的译文出现。

一、名词的转译

英语中名词的转译就是把英语中的名词转化为其他的词性形式，包括名词转译为动词、名词转译为形容词、名词转译为副词等。

1. 名词转译为动词

英语词汇有很多属于同根词，也就是说一个词根可以派生出很多词性的词来，如英语中抽象名词、动名词、由动词本身派生出来的名词、由介词构成的介词短语等，翻译时可以适当加以变通处理。例如：

(1) In some cases, deserts are the *creation* of *destruction* of virgin forest.

译文：沙漠有时是人类毁坏原始森林造成的。（creation 和 destruction 作为动词派生的名词被转译为动词"造成"和"毁坏"）

(2) *With the use of* liquid crystal display, TV set can be made small.

译文：利用液晶显示，电视机可制得很小。（介词名词短语 with the use of 被转译为动词"利用"）

(3) In the dynamo, mechanical energy is used for *rotating* the armature between the poles of an electromagnet

译文：在直流发电机中，机械能用来转动电磁铁两极之间的电枢。（动名词 rotating 被转译为动词"转动"）

2. 名词转译为形容词

一些做表语的形容词或由形容词派生的名词，在翻译的过程中往往会被转译为形容词，例如：

(1) Single crystals of high perfection are an absolute *necessity* for the fabrication of integrated circuits.

译文：高度完整的单晶对于制造集成电路来说是绝对必要的。（由 necessary 派生的名词 necessity 译为形容词"必要的"）

(2) Without the gas pipelines, the movement of large volumes of gases over great distances would be an economic *impossibility*.

译文：如果没有煤气管道，远距离输送大量煤气在经济上就是不可行的。（由形容词派生的名词 impossibility 转译为形容词"不可行的"）

3. 名词转译为副词/连接词（组）

在有些语境中，名词必须转译为副词，句子才显得通畅。例如：

(1) The advantage of ion implantation is the *uniformity* of doping

译文：离子注入的优点是掺杂均匀。（由形容词派生的名词 uniformity 被转译为副词"均匀"）

(2) All the engineers in the company find *difficulty* in solving this problem.

译文：公司所有的工程师都觉得很难解决这个问题。（名词 difficulty 转译为副词"很难"）

二、动词的转译

1. 动词转译为名词

英语中动词和名词之间的转换非常普遍，很多英语名词都是由动词派生而来的。在翻译过程中，为了满足表达需要，有时也会将动词转译为名词。例如：

A highly developed physical science *is characterized* by an extensive use of mathematics.

译文：一门高度发展的自然科学的特点是广泛地运用数学。（动词 be characterized 被转译为名词"特点"）

2. 动词转译为形容词

动词有时也转译为形容词。例如：

(1) The output voltages of the control system *varied* in a wide range.

译文：这台控制系统输出电压的变化范围很宽。（动词 vary 被转译为形容词"宽"）

(2) Tests showed that the computer *operated* at a speed of at least fifty million cycles/second.

译文：试验结果表明，这台计算机工作的速度不低于 5000 万次/秒。（动词 operate 被转译为形容词"工作的"）

3. 动词转译为副词/连接词（组）

动词转译为副词不太多见，翻译的时候需要多加注意。例如：

(1) Rapid evaporation at the heating-surface tends to make the stream wet.

译文：加热面上的迅速蒸发，往往使蒸汽的湿度变大。（动词 tend 转译为副词"往往"）

(2) As relations between China and Australia develop, the continuing importance of expanding trade will *be balanced by* the development of close contact over a broad range of political issues.

译文：随着中澳两国关系的发展，扩大贸易仍将是重要的，相应地还要在一系列广泛政治问题上展开密切的联系。（动词 be balanced by 转译为副词连接词组"相应地还要在……上"）

三、形容词的转译

英语中的形容词也可以根据不同情况转译为其他汉语词性，如将形容词转译为动词、形容词转译为名词、形容词转译为副词等。

1. 形容词转译为动词

英语中用于修饰名词的形容词很多，也有很多是和系动词构成的系表结构表示情感、心态、知觉、态度等，往往需要改变词性译为动词。例如:

(1) If extremely low-cost power were ever to become *available* from large nuclear power plants, electrolytic hydrogen would become competitive.

译文：如果能够从大型核电站获得成本极低的电力，电解氢会更具竞争力。（形容词

available 转译为动词"能够"）

(2) Scientists are *confident* that all matter is indestructible.

译文：科学家们深信一切物质都是不灭的。（形容词 confident 转译为动词"深信"）

2. 形容词转译为名词

充当表语、补足语等其他成分的形容词，在必要的时候也可以转译为名词。例如：

(1) Glass is more *transparent* than plastic cloth.

译文：玻璃的透明度比塑料布高。（形容词 transparent 转译为名词"透明度"）

(2) The nature of the organism causing measles remained *unclear* until 1911, when it was proved to be a virus.

译文：引起麻疹的有机体的性质曾是一个不解之谜，直到1911年才被证实是一种病毒。（形容词 unclear 转译为名词"不解之谜"）

3. 形容词转译为副词

当英语名词转译为动词时，修饰它的形容词就相应地转译为副词。例如：

(1) The *wide* application of electronic machines in scientific work, in designing and in economic calculations will free man from the labor of complicated computations.

译文：在科学研究、设计和经济计算等方面广泛地应用电子计算机，可以将人们从繁重的计算工作中解放出来。（形容词 wide 转译为副词"广泛地"）

(2) With *slight* modification each type can be used for all three system.

译文：每种型号只要稍微改动就能用于这三种系统。（形容词 slight 转译为副词"稍微"）

四、副词的转译

英语中副词的使用非常广泛，翻译时，可以根据具体需要做一些调整和转译，如副词转译为动词、副词转译为名词、副词转译为形容词等。

1. 副词转译为动词

英语中副词在修饰表语、补语或复合宾语的时候具有动词的意味，表示一定的动作或方向，这类副词往往需要转译。此外，还有一些和系动词构成的合成谓语，或做宾语补足语，或做状语，也需要转译为动词。例如：

(1) When the switch is off, the circuit is open and electricity doesn't go *through*.

译文：当开关断开时，电路形成开路，电流就不能通过。（副词 through 译为动词"通过"）

(2) We finished the project two days *ahead* of schedule.

译文：我们比原计划提前两天完成了该项目。（副词 ahead 译为动词"提前"）

2. 副词转译为名词

英语中许多副词是由动词或形容词词根派生而来，对于这些副词习惯上还是转译为名词。例如：

(1) Sodium is very active *chemically*.

译文：金属钠的化学性质很活跃。（副词 chemically 转译为名词"化学性质"）

(2) All structural materials behave *plastically* above their elastic range.

译文：超过弹性极限时，一切结构材料都会显示出塑性。（副词 plastically 转译为名词"塑性"）

3. 副词转译为形容词

英语中有些副词是形容词派生而来的，在翻译过程中，通常会把副词还原为形容词。例如：

(1) The resistivity of semiconductors is *inversely* proportional to temperature.

译文：半导体的电阻率与温度成反比。（副词 inversely 转译为形容词"相反的"）

(2) Neutrons act *differently* from protons.

译文：中子的作用不同于质子。（副词 differently 转译为形容词"不同的"）

五、介词的转译

介词是英语中最常见的一种词，不仅数量多，而且在使用范围和使用频率上也超过其他词性的词。相比之下，汉语的介词没有那么多，使用也没有那么广。因此在翻译过程中，介词就需要加倍小心。英语里的介词具有较强的动词意味，通常情况下，都可以转译为动词。例如：

(1) Except for atomic energy, all forms of energy used by man are *from* the sun.

译文：除原子能之外，人类所利用的一切形式的能量都来自太阳。（介词 from 转译为动词"来自"）

(2) There are many substances *through* which electric currents will not pass at all.

译文：有许多物质，电流是根本不能通过的。（介词 through 转译为动词"通过"）

六、代词及其他词性的转译

为了准确表达原文的意思，代词及其他词性在翻译时也需要转译。例如：

(1) A decision-making framework is likely to be a series of relatively simple steps, or a process, *that* represents the strategy, policy, and related issues.

译文：决策框架很可能是一系列相对简单的步骤，或某个过程；这个过程代表战略、政策及有关问题。（代词 that 转译为名词短语"这个过程"）

(2) *The* chariot can be driven at 200 miles an hour.

译文：这种战车的速度可达每小时 200 英里。（冠词 the 转译为指示代词"这种"）

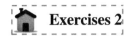 **Exercises 2**

Translate the following paragraph into Chinese.

(1) The management process begins with an understanding of the definition and classification of the different wastes, and their harmful effects on human health, and ends with the application of risk management to control human health and environmental impacts of hazardous waste.

(2) Even though the contents of the canal were highly toxic, it took a while before health problems were noticed and residents complained.

(3) Some of these emissions are potentially harmful to human health and environment, and thus warrant special techniques of management.

(4) The motivation to manage hazardous wastes to minimize human health and environmental risks came from a public awareness that grew over many historical incidents of uncontrolled disposal of hazardous waste.

Part 3　Writing

学术论文的撰写（六）——学术论文前言的编写

科技英语论文最重要的内容部分通常由前言、正文和结论组成。现在，我们先来看看前言部分的内容和写作方法。

前言（introduction），又称为引言。在长篇论文中也称为导论、导言、绪言、绪论等。如果说摘要是全篇论文的缩影，那么引言则是科技论文的帽子，它向读者初步介绍文章内容，解释文章的主题、目的和总纲。短篇论文通常在正文之前用简洁语言略述总纲作为前言，而对长篇论文和涉及某些不常见的内容的论文来说，前言需要更加翔实，以便于读者阅读文章，引导读者理解科学成果的意义、实验采用的方法和论文展开的计划等。

前言一般至少应包括以下三方面的内容：

（1）说明论文的主题和目的。

（2）说明论文写作的情况及背景。

（3）概述达到理想答案的方法。

需要指出的是，前言不应重述摘要或解释摘要，不应对实验的理论、方法和结果做详尽叙述，也不应提前使用结论和建议。在前言中引用的参考文献要仔细地加以选择，以便提供最重要、最适宜的背景材料。

同时，写作前言时还应根据读者的特点和需要来决定写作基调。需要对哪些专业术语进行明确的定义，而哪些术语是学科基础无需多做解释的，哪些背景和目的需要详细阐述，而哪些是可以简洁概述的？总之，写作应做到有的放矢。

在前言中，作者不需要对自己的研究工作和自己的能力表示谦虚，让读者对论文做出自己的评价。

前言可以由一段文字组成，也可由几段文字组成，这主要取决于论文的长短。在有的前言之前，作者用"introduction"来标明，而在有的前言（如只是一段文字的前言）之前往往没有使用"introduction"来标明。下面介绍 5 种前言的写作方法。

一、直述主题开头法

直述主题开头法就是在引言的第一句话中，直接点透论文的主题，然后再开始叙述其他因素。

【例 1】 Writing is one of the most difficult tasks for language learners. It is very much the case in learning English. Many students who have been learning English for many years are very poor in writing at the sentence level. Some of the compositions are often even not revisable, because the sentences written do not make any sense. On the other hand, few students can write well in English. Those who can write English well usually have good study habits, and good reading-habits in particular. They pay much more attention to ideas while reading.

Language-learning is a thinking process. Thinking is involved in every task of learning from reading to speaking, and especially writing. Good writing needs creative thinking. So to develop students' thinking is very important in helping students to write, And this should be done through the process of learning. This paper discusses four areas of suggestions on developing students' thinking.

例 1 中，前言的第一句话就介绍了论文的主题，然后围绕"写作是语言学习者最困难的任务之一"介绍了本文的主要内容，对写作感到困难的原因进行了分析，并就此提出了如何发展学生的思维能力，为论文的全面展开做了铺垫。这是一个典型的使用直叙主题开头的前言，这种方法最为常用。

二、利用一个问题或一系列问题开头的方法

在前言写作时，为了引起读者的注意，采用一个问题或一系列问题开头不失为一种十分有效的方法。当读者读到问题时，就会急切地想知道问题的答案，给人一种悬念。例如在一篇论文中，其前言是用这样一个疑问句开头的："How much bureaucratic stupidity do we have put up with?"当读者一读到这个问题就会产生"有多少官僚主义的愚蠢行为？是哪些？我们为什么会容忍？"等一系列有待于从论文中寻找答案的问题。这样容易吸引读者，也便于读者了解论文的内容和进行检索时决定取舍。例 2 是一个以问题开头的前言。

【例 2】 What are the chances of a nuclear war in the near future? How many Americans would survive a nuclear attack? Would such an attack make living conditions impossible for the survivors? These and similar questions are being asked by citizens' groups throughout this country as they debate the issue of arms control and nuclear disarmament.

例 2 中，一开始作者用三个问题提出了全美国人民所关心核战的可能性，有多少人可

以幸免于难，幸存者将来的生活条件是否存在。紧接问题之后，涉及了论文的主题——核裁军与军备控制，为论文对这一问题的讨论打下了基础。要注意的是，所提的问题必须在论文中进行解答，同时所提的问题必须是文章中的内容。千万不要写出一些与文章主题关系不大或者根本没有关系的问题。

三、引语开头法

利用引语开头是点明论文主题的一种有效的方法，引语开头容易引起读者的注意。例如在一篇论述教育方式改革多样性的论文中首先引用了《论语》中的 "Make no social distinctions in teaching（有教无类）"，提出教育应面向全社会，因此教育方式也应是多种多样的，为了表示引语的权威性，在这句话后紧接着作者就写了一句 "This is a quotation from the Analects from Confucius who was the greatest educator in China"，这句话不仅说明了引语的出处，而且因为《论语》是广为人知的经典，这样就充分体现了引语的权威性。同时，也为论文围绕这一主题来展开讨论做好了铺垫并提供了强有力的依据。所以，在使用引语的时候，所引的话最好是有一定权威性的著作、名人的名言，或某一技术领域、专业权威人士的文字或语言。下面是一个引用著名医学家 Bernard Baruch 的话为开头的论文的前言。

【例3】 "There are no such things as incurables," said Bernard Baruch, "there are only things for which man has not found a cure." A lifetime in medical practice, education and research has convinced me, too, that we need not accept disease as an inescapable human destiny, despite our lack of information about many forms of human illness.

例3在一开头就引用了著名医学家 Bernard Baruch 关于治疗疾病的论断，他提出了应正确对待疾病，从中就可以看出这篇论文的主题是关于如何认识疾病和正确对待疾病的。

四、定义开头法

用定义对文中所要论述的主题开头有利于作者表达，这样既可以点明论文的主题又可以对文中将要论述的中心问题进行必要的解释。当然有时定义单用一句话解释不全，这时就要使用一段文字对其进行定义。例4就是一个使用一段文字定义为引言的实例。

【例4】 A nurse is one who looks after patients. He/She helps the doctor, who cures sick persons by writing out prescriptions. And it is the nurse with whom the patients spend most of the time in a ward room.

例4是一个用定义"护士的职责和工作"为引言的典型实例。它首先解释了护士工作总的概念，然后叙述了护士的工作和范围，并且用了一个非限制性定语从句非常巧妙地将护士的工作范围与医生的工作范围做了比较。从这个引言不难看出论文的主题将是叙述护士应怎样履行自己的职责。

五、利用统计数字开头法

利用统计数字开头的引言是为了利用数字来说明某一问题，其目的是说明文章主题的重要性，为在论文中对数据的采集方法、使用数据的目的进行论述前的准备工作。下面是一个以数据统计为开头的实例。

【例 5】 The fact that less than 5% of the British population graduate from universities may seem surprising, especially when compared with American percentage of over 30%.

例 5 中可以看出论文的主题是对英美两国人口中大学毕业生所占的比例进行比较，从中找出形成这一对比数字的某些差异。

前言写作的关键要在"引"上下功夫，从上述介绍的 5 种方法可以看出，前言除了它本身应讲述的内容外，主要还在于为正文的铺垫下功夫，为正文的内容起到一个引导的作用。作为前言的结束需要指出的是，如果前言是对某一自然学科的讨论，往往使用第三人称单数，其时态根据需要使用一般现在时或过去时即可。如果前言是对某一人文社会科学问题的讨论，可使用其他人称。

下面提供一篇前言范文，供写作时参考。

【范文 1】

The collection and treatment of domestic sewage and wastewater is vital to public health and clean water. It is among the most important factors responsible for the general level of good health enjoyed in the United States. Sewers collect sewage and wastewater from homes, businesses, and industries and deliver it to wastewater treatment facilities before it is discharged to water bodies or land, or reused.

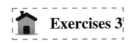 **Exercises 3**

Have you written a scientific article? Now you can write one. After learning the skills of writing introduction, you can prepare to write a scientific article about your daily discharge of MSW, and then write one or two paragraphs of introduction.

Preventing Trash at the Source

Overview

You must make sure hazardous waste produced or handled by your business in England causes no harm or damage.

You have responsibilities known as your "duty of care". You must also meet extra requirements depending on whether you're a waste:

◆ producer or holder （you produce or store waste）

- carrier (you collect and transport waste)
- consignee (you receive waste, such as for recycling or disposal)
- Check what you need to do in Northern Ireland, Scotland and Wales.
- There are different requirements for exporting waste.
- Check if your waste is hazardous.

Waste is generally considered hazardous if it (or the material or substances it contains) is harmful to humans or the environment. Examples of hazardous waste include:

- asbestos
- chemicals, such as brake fluid or print toner
- batteries
- solvents
- pesticides
- oils (except edible ones), such as car oil
- equipment containing ozone depleting substances, like fridges
- hazardous waste containers

Producers and holders

You must follow these steps in England if your business:

- produces hazardous waste
- holds or stores hazardous waste
- has hazardous waste removed from its premises

Classify your waste to check if it's hazardous. Separate and store hazardous waste safely. Use authorised businesses to collect, recycle or dispose of your hazardous waste—check that waste carriers are registered and waste sites have environmental permits.

Carriers

You must follow these steps if your business collects and transports hazardous waste in England (for example you're a waste carrier, or you move your own waste).

- Register as a waste carrier.
- Check parts A and B of the consignment note and the waste before you accept it—make sure the waste is classified correctly.
- Separate waste correctly when you load it for transportation.
- Fill in the part of the consignment note that applies to you.
- Leave one copy of the consignment note with the waste producer or holder and keep 2 copies—these must stay with the waste until it reaches its destination.
- Take the waste to the destination on the consignment note — it must be an authorised waste site.

◆ Keep records (known as a "register") for one year. You must keep records at your head office.

Consignees

You must follow these steps if you receive, treat or dispose of hazardous waste at premises in England.

◆ Get an environmental permit or register an exemption for your premises.

◆ Check the consignment note and waste before you accept it — make sure it's classified correctly.

◆ Reject the waste if the consignment note is missing, incorrect or incomplete.

◆ Fill in part E of the consignment note for any hazardous waste you accept or reject —keep one copy and hand one copy back to the carrier.

◆ Send consignee returns to the Environment Agency, and the waste producer or holder, to report on any hazardous waste you accept or reject.

◆ Keep records (known as a "register"). You must keep records at the site where the hazardous waste was stored, treated or disposed.

(*https://www.gov.uk/dispose-hazardous-waste*)

Unit 4 Solid Wastes and Soil Pollution

Lesson 3 Soil Pollution

Part 1 Reading

1 Definition

Soil pollution is defined as the build-up in soils of persistent toxic compounds, chemicals, salts, radioactive materials, or disease causing agents, which have adverse effects on plant growth and animal health.

Soil is the thin layer of organic and inorganic materials that covers the Earth's rocky surface. The organic portion, which is derived from the decayed remains of plants and animals, is concentrated in the dark uppermost topsoil. The inorganic portion made up of rock fragments, was formed over thousands of years by physical and chemical weathering of bedrock. Productive soils are necessary for agriculture to supply the world with sufficient food.

There are many different ways that soil can become polluted, such as:

(1) Seepage from a landfill.

(2) Discharge of industrial waste into the soil.

(3) Percolation of contaminated water into the soil.

(4) Rupture of underground storage tanks.

(5) Excess application of pesticides, herbicides or fertilizer.

(6) Solid waste seepage.

The most common chemicals involved in causing soil pollution are:

(1) Petroleum hydrocarbons.

(2) Heavy metals.

(3) Pesticides.

(4) Solvents.

2 Types of Soil Pollution

(1) Agricultural Soil Pollution.

① Pollution of surface soil.

② Pollution of underground soil.

(2) Soil pollution by industrial effluents and solid waste.

① Pollution of surface soil.

② Disturbances in soil profile.

(3) Pollution due to urban activities.

① Pollution of surface soil.

② Pollution of underground soil.

3 Causes of Soil Pollution

Soil pollution is caused by the presence of man-made chemicals or other alteration in the natural soil environment. This type of contamination typically arises from the rupture of underground storage links, application of pesticides, percolation of contaminated surface water to subsurface strata, oil and fuel dumping, leaching of wastes from landfills or direct discharge of industrial wastes to the soil. The most common chemicals involved are petroleum hydrocarbons solvents, pesticides, lead and other heavy metals. This occurrence of this phenomenon is correlated with the degree of industrialization and intensities of chemical usage.

A soil pollutant is any factor which deteriorates the quality, texture and mineral content of the soil or which disturbs the biological balance of the organisms in the soil. Pollution in soil has adverse effect on plant growth.

Pollution in soil is associated with:
- Indiscriminate use of fertilizer.
- Indiscriminate use of pesticides, insecticides and herbicide.
- Dumping of large quantities of solid waste.
- Deforestation and soil erosion.

4 Indiscriminate Use of Fertilizers

Soil nutrients are important for plant growth and development. Plants obtain carbon, hydrogen and oxygen from air and water. But other necessary nutrients like nitrogen, phosphorus, potassium, calcium, magnesium, sulfur and more must be obtain from the soil. Farmers generally use fertilizer to correct soil deficiencies. Fertilizers contaminate the soil with impurities, which come from the raw materials used for their manufacture. Mixed fertilizers often contain ammonium nitrate (NH_4NO_3), phosphorus as P_2O_5, and potassium as K_2O. For instance, As, Pb and Cd present in traces in rock phosphate mineral get transferred to super phosphate fertilizer. Since the metals are not degradable, their accumulation in the soil above their toxic levels due to excessive use of phosphate fertilizers, becomes an indestructible poison for crops.

5 Indiscriminate Use of Pesticides, Insecticides and Herbicides

Plants on which we depend for food are under attack from insects, fungi, bacteria, viruses, rodents and other animals, and must compete with weeds for nutrients. To kill unwanted populations living in or on their crops, farmers use pesticides. The first widespread insecticide use began at the end of World War II and included DDT (dichlorodiphenyltrichloroethane). Insects soon became resistant to DDT and as the chemical did not decompose readily, it persisted in the environment. Since it was soluble in fat rather than water, it biomagnified up the food chain and disrupted calcium metabolism in birds, causing eggshells to be thin and fragile. As a result, large birds of prey such as the brown pelican, ospreys, falcons and eagles became endangered. DDT has been now been banned in most western countries. Ironically many of them including U.S.A., still produce DDT for export to other developing nations whose needs outweigh the problems caused by it.

Pesticides not only bring toxic effect on human and animals but also decrease the fertility of the soil. Some of the pesticides are quite stable and their biodegradation may take weeks and even months.

Pesticide problems such as resistance, resurgence, and health effects have caused scientists to seek alternatives. Pheromones and hormones to attract or repel insects and using natural enemies or sterilization by radiation have been suggested.

6 Dumping of Solid Wastes

In general, solid waste includes garbage, domestic refuse and discarded solid materials such as those from commercial, industrial and agricultural operations. They contain increasing amounts of cardboards, plastics, glass, old construction material, packaging material and toxic or otherwise hazardous substances. Since a significant amount of urban solid waste tends to be paper and food waste, the majority is recyclable or biodegradable in landfills. Similarly, most agricultural waste is recycled and mining waste is left on site.

The portion of solid waste that is hazardous such as oils, battery metals, heavy metals from smelting industries and organic solvents are the ones we have to pay particular attention to. These can in the long run, get deposited to the soils of the surrounding area and pollute them by altering their chemical and biological properties. They also contaminate drinking water aquifer sources. More than 90% of hazardous waste is produced by chemical, petroleum and metal-related industries and small businesses such as dry cleaners and gas stations contribute as well.

7 Deforestation

Soil Erosion occurs when the weathered soil particles are dislodged and carried away by wind or water, Deforestation, agricultural development, temperature extremes, precipitation including acid rain, and human activities contribute to this erosion. Humans speed up this process by construction, mining, cutting of timber, over cropping and overgrazing. It results in floods and cause soil erosion.

Forests and grasslands are an excellent binding material that keeps the soil intact and healthy. They support many habitats and ecosystems, which provide innumerable feeding pathways or food chains to all species. Their loss would threaten food chains and the survival of many species. During the past few years quite a lot of vast green land has been converted into deserts. The precious rain forest habitats of South America, tropical Asia and Africa are coming under pressure of population growth and development (especially timber, construction and agriculture). Many scientists believe that a wealth of medicinal substances including a cure for cancer and aids, lie in forests. Deforestation is slowly destroying the most productive flora and fauna areas in the world, which also form vast tracts of a very valuable sink for CO_2.

8 Pollution Due to Urbanization

Urban activities generate large quantities of city wastes including several biodegradable materials (like vegetables, animal wastes, papers, wooden pieces, carcasses, plant twigs, leaves, cloth wastes as well as sweepings) and many non-biodegradable materials (such as plastic bags, plastic bottles, plastic wastes, glass bottles, glass pieces, stone/cement pieces).

9 Environmental Long Term Effects of Soil Pollution

If contaminated soil is used to grow food, the land will usually produce lower yields than it would if it were not contaminated. This, in turn, can cause even more harm because a lack of plants on the soil will cause more erosion, spreading the contaminants onto land that might not have been tainted before.

In addition, the pollutants will change the makeup of the soil and the types of microorganisms that will live in it. If certain organisms die off in the area, the larger predator animals will also have to move away or die because they've lost their food supply. Thus, it's possible for soil pollution to change whole ecosystem.

10 Effects of Soil Pollution in Brief

- Pollution runs off into rivers and kills the fish, plants and other aquatic life.
- Crops and fodder grown on polluted soil may pass the pollutants on to the consumers.
- Polluted soil may no longer grow crops and fodder.
- Soil structure is damaged (clay ionic structure impaired).
- Corrosion of foundations and pipelines.
- Impairs soil stability.

- May release vapors and hydrocarbon into buildings and cellars.
- May create toxic dusts.
- May poison children playing in the area.

(http://www.index-files.com/file-pdf/soil-pollution-pdf)

Words and Expressions

uppermost	[ˈʌpərmoʊst]	adj. 最高的，最上端的，最上面的，最重要，最关键
bedrock	[bedrɑːk]	n. 基岩（松软的沙、土层下的岩石）
seepage	[ˈsiːpɪdʒ]	n. 渗，渗透，渗液
rupture	[ˈrʌptʃər]	n. （体内组织等的）断裂、破裂，断裂，爆裂，疝气； v. （使体内组织等）断裂、裂开、破裂，（使容器或管道等）断裂
petroleum hydrocarbons		n. 石油烃
solvent	[ˈsɑːlvənt]	n. 溶解，溶剂，有机溶剂
strata	[streɪtə]	n. 地层
indiscriminate	[ˌɪndɪˈskrɪmɪnət]	adj. 无差别的，不加选择的
potassium	[pəˈtæsiəm]	n. 钾
ammonium nitrate		n. 硝铵，硝酸铵
fungi	[ˈfʌndʒaɪ]	n. 真菌，fungus 的复数
rodent	[ˈroʊdnt]	n. 啮齿动物
DDT		n. 双对氯苯基三氯乙烷
disrupt	[dɪsˈrʌpt]	v. 扰乱，使中断，打乱
metabolism	[məˈtæbəlɪzəm]	n. 新陈代谢
brown pelican		n. 褐鹈鹕
osprey	[ˈɔsprɪz]	n. 鹗，鱼鹰
falcon	[ˈfælkənz]	n. 隼
resurgence	[rɪˈsɜːrdʒəns]	n. 复苏，复兴
pheromone	[ˈferəmoʊn]	n. 外激素，信息素
hormone	[ˈhɔːrmoʊn]	n. 激素，荷尔蒙
repel	[rɪˈpel]	v. 击退，驱逐，推开，赶走，驱除
sterilization	[ˌsterəlaɪˈzeɪʃ(ə)n]	n. 灭菌，绝育
dislodge	[dɪsˈlɑːdʒ]	v. （把某物）强行去除，取出，移动
intact	[ɪnˈtækt]	adj. 完好无损，完整

timber	['tɪmbər]	n. （建筑等用的）木材，木料
flora	['flɔːrə]	n. （某地区、环境或时期的）植物群
fauna	['fɔːnə]	n. （某地区或某时期的）动物群
carcasses	['kɑrkəsɪz]	n. 动物尸体
twig	[twɪɡ]	n. 细枝，小枝
sweeping	['swiːpɪŋ]	n. 扫除，（pl.）垃圾；
		v. 堆积（东西），堆置
taint	[teɪnt]	v. 使腐坏，污染，玷污、败坏（名声）
fodder	['fɑːdər]	n. （马等家畜的）饲料，秣
cellar	['selər]	n. 地窖，地下室

Exercises 1

1. According to the reading material, chose the best answer(s) from the options.

(1) Soil pollution is defined as _____ .

A. the build-up in soils of persistent toxic compounds, chemicals, salts, radioactive materials, or disease causing agents, which have adverse effects on plant growth and animal health.

B. the thin layer of organic and inorganic materials that covers the Earth's rocky surface.

C. the presence of man-made chemicals or other alteration in the natural soil environment.

D. none of them.

(2) The most common chemicals involved in causing soil pollution are _____ .

A. Petroleum hydrocarbons   B. Heavy metals   C. Pesticides   D. Solvents

(3) Types of soil pollution are _____ .

A. municipal sewage

B. industrial effluents and solid waste

C. pollution due to urban activities

D. agricultural Soil Pollution

(4) Pollution in soil is associated with _____ .

A. indiscriminate use of pesticides, insecticides and herbicide

B. deforestation and soil erosion

C. dumping of large quantities of solid waste

D. indiscriminate use of fertilizer

(5) Effects of soil pollution are _____ .

A. creating toxic dusts   B. reducing the crop yield

C. releasing vapors and hydrocarbon   D. damaging soil structure

2. Describe soil erosion process in English.

3. Describe the impacts of soil pollution to the ecosystem in English.

Part 2　Translation

英语翻译中长句的翻译

英语句子具有长短之分，一般的长句可以分为简单句、并列句或复合句。方梦之教授认为"长句指的是词数多、结构复杂的句子。英语长句一般适用于精确、周密、细致地描述事物间的复杂关系，多用于书面语体，如政论、科技语体"。

科技英文中长句的翻译是翻译实践中最令人犯难的问题。说其难，是因为这类句子往往还包含了多个短语或是从句，一层套一层，使句子结构变得更加复杂，难以理解。例如：

(1) Many man-made substances are replacing certain natural materials because either the quantity of the natural product can't meet our ever-increasing requirement, or, more often, because the physical properties of the synthetic substance, which is the common name for man-made materials, have been chosen, and even emphasized, so that it would be of the greatest use in the fields in which it is to be applied.

译文：人造材料通称为合成材料。许多人造材料正在代替某些天然材料，这是由于天然产物的数量不能满足日益增长的需要，或者是由于人们选择了合成材料的一些物理性质并加以突出，因此，合成材料在拟用的领域中将具有极大的用途。

英语中，句子结构可以借助各种连接手段加以扩展和组合，形成纷繁复杂的长句，故而人们喜欢使用成分（比如主句和短语等）嵌套的长句；而在汉语中，少用甚至不用连接词语，因而语段结构流散，但语意层次分明，故而人们喜欢使用独立句和较短的单句，句子之间的关系主要由上下文语义来连贯。对此现象，我国著名语言学家王力先生一针见血地指出："西洋语的结构好像连环，虽然环与环都联络起来，毕竟有联络的痕迹；中国语的结构好像无缝天衣，是一块块地硬凑，凑起来还不让它有痕迹。西洋语法是硬的，没有弹性的；中国语法是软的，富于弹性的。"

值得注意的是，英语中的长句都是由一些基本的句型扩展或变化而来的，也就是说在以"主+谓"为主干的英语句子中，添加各种关系词，把有关的材料组成其枝干，使句子丰腴起来。《英语基础语法新编》中指出："英语各种长短句子，一般都可以看作是这五种基本句型及其变式、扩展、组合、省略或倒装。"因此，翻译长句的关键首先在于抓住全句的中心内容，弄清句子的组织构成以及相互间的语法关系，尤其是单词、短语、从句的并列层次和扩展层次。例如：

(2) Now the integrated circuit has reduced by many times the size of the computer of which it forms a part, thus creating a new generation of portable minicomputers.

译文：现在集成电路成为计算机的组成部分，使计算机的体积大大缩小，从而产生了新一代的袖珍式微型计算机。

本句属于基本句型的扩展。原句中，句子的主体结构为："…circuit has reduced…the size…"。只是为了逻辑或者是修辞的需要，增加了定语和状语等修饰语。当然，一般情况下，这些修饰语可以是单词或词组，也可以是从句。又如：

(3) The structure of the steel and the resulting properties will depend on how hot the steel gets and how quickly or slowly it is cooled.

译文：钢的结构及其形成的性质，取决于加热的温度和冷却的速度。

本句属于基本句型成分的扩展。在原句中，句子基本句型的成分是单词，即"will depend on…"，在增加了词组和从句之后，句型结构得到了扩展。

一般来说，就翻译程序而言，长句翻译可以按六步进行操作，即

①紧缩主干（找出主句主语、谓语及宾语）；

②区分主从（理清主从句，找出从句的主语、谓语及宾语以及修饰关系）；

③辨析词义（特别是主干中的词义）；

④分清层次（推断出句子思维逻辑发展形势及重心）；

⑤调整搭配（特别是主谓及动宾搭配，调整语序）；

⑥润饰词语语句（考虑选词炼句并考虑文体的适应性）。

以上六步看似复杂，其实与其他翻译过程一样，必须经过三个阶段：一是对原文的准确理解；二是对译文的恰当表达；三是对译文的适当润色和修改。否则，翻译难以通达。例如：

(4) Having achieved this position the faceplate should be held to prevent rotation the timing shaft should be now depressed to disengage the drive and rotated clockwise until the line "B"（fig. 12）on the cam shaft below coincides with the line on the information plate.

译文1：当达到这一位置后，应当保持面板以防旋转，直到凸轮轴下的线"B"（图12）与铭牌线路相重叠时，定时轴应降低，以便传动装置脱扣，而按顺时针方向转动。

译文2：在达到这一位置时，应当握住面盘防止转动；压下定时轴，使传动机构松脱；并逆时针方向旋转定时轴直到凸轮轴下部"B"线（图12）与指示盘上刻线重合。

比较译文1和译文2，不难发现其间的差异。原文的最大特点是用一系列并列句，顺次交代一系列的操作过程，"until"只是最后一句谓语"be rotated"的限定从句，如果能把这层关系摸透，便可得到译文2这样的佳译。

应该说，原文没有太复杂的句法结构，一气呵成，叙说明晰，无半点含糊。译文1没有理解三个句子之间的逻辑联系，没有将"until"从句对号入座，没能理解"depressed"和"rotated"同是"shaft"的谓语动词。在翻译过程中，原文中明晰的逻辑关系已丧失殆尽。

虽然英语基本语法并不是太难，但在具体运用过程中，句子形态和长短千变万化，难以归纳为一些简单的模式。要想攻破长句翻译这座堡垒，译者需要扎实的中英文功底和广博的知识。以下简要探讨几种长句的常见翻译方法：

一、顺序译法

顺序译法指按照原句的顺序进行翻译，这不仅可以避免译文产生混乱，还可以维持原文

结构。英语中，有些长句所叙述的一连串动作基本上是动作发生的先后顺序，也有些长句的内容是按逻辑关系安排的，这些与汉语里表达方法一致，故可用顺序译法。此外，一些并列复合句和带宾语从句或表语从句的主从复合句，通常可按照原文的思路和顺序来翻译。例如：

(1) Nations will usually produce and export those goods in which they have the greatest comparative advantage, and import those items in which they have the least comparative advantage.

译文：各国通常都生产和出口那些他们最具优势的产品，而进口那些他们最不具优势的产品。

原句中，使用了两个由"in which"引导的定语从句，在译文中分别被译成"他们最具优势的"和"他们最不具优势的"，行文简洁精练，两个"in which"都被省略了，译文采用的是顺序译法。

(2) We know from the fossil record that our ancestors and other intelligent creatures and australopithecines, branched off from an apelike creature 2.5 million to 3 million years ago, and coexisted until the australopithecines died out a little less than a million years ago.

译文：我们从化石记录得知，我们的祖先和另一类有智慧的生物——南方古猿——在250万~300万年前从猿类分化出来并且共同生存，直到将近100万年前南方古猿灭绝为止。

原文虽然较长，但结构相对比较简单，可按意义将其分为三个意群，逻辑关系及时间顺序与汉语基本一致，故顺译即可。

二、逆序译法

英语中，有些句子的表达顺序与汉语的表达习惯不同，甚至相反，尤其是一些复合句，其主句一般放在句首，即重心在前，而汉语则一般按时间和逻辑顺序，将主要部分放在句尾，形成尾重心。对于这些句子，翻译时应当采用逆序翻译法。例如：

(1) The time is now near at hand which must probably determine whether Americans are to be freemen or slaves; whether they are to have any property they can call their own; whether their houses and farms are to be pillaged and destroyed, and themselves consigned to a state of wretchedness from which no human efforts will deliver them.

译文：美国人将会成为自由人还是奴隶，他们是否将拥有称得上属于自己的财产，他们的房屋和农庄是否会遭到掠夺和毁坏，他们自己是否会陷入任何人力都无法拯救的悲惨境地——现在，必须决定这一切的时刻就在眼前。

原文中，行文方式是先总说，即先说出结论，接着再谈具体事情；而译文则先谈论具体情况，最后再分说，这也是英汉两种语言在行文方式上的不同。造成这种行文方式不同的主要原因是英美人的思维方式与中国人不同，表现在语言上就存在这种差异。

(2) It therefore becomes more and more important that, if students are not to waste their opportunities, there will have to be much more detailed information about courses and more advice.

译文：因此，如果要使学生充分利用他们（上大学）的机会，就得为他们提供大量关于课程的更为详尽的信息，做更多的指导。这个问题显得越来越重要了。

原文是由一个主句、一个条件状语从句和一个宾语从句组成，"……变得越来越重要"是主句，也是全句的中心内容，全句共有三个谓语结构，包含三层含义：①……变得越来越重要；②如果要使学生充分利用他们的机会；③得为他们提供大量更为详尽的信息，做更多的指导。为了使译文符合汉语表达习惯，翻译时采用逆序翻译法。

三、分译译法

分译是指原文语句过长，同时又含有多层意思，而主句和从句或者修饰部分之间的关系又不是很紧密，如果在译文中用一个句子对原文进行翻译，反而会引起句意混乱，翻译时可按汉语多用短句的习惯，把长句中的从句或短语化为句子，分开进行叙述。一般来说，分译的目的是使句子意思的表达清楚明了。例如：

(1) The loads that a structure is subjected to are divided into dead loads, which include weights of all the parts of the structure, and live loads, which are due to the weights of people, movable equipment, etc.

译文：一个结构物受到的荷载可分为静载与动载两类。静载包括该结构物各部分的重量。动载则是由人及可移动设备等的重量而引起的荷载。

原文结构比较清晰，翻译时要做一些调整。新手很可能会将其译为"结构物受到的荷载可分为包括该结构物各部分重量的静载和由于人及可移动设备等的重量而引起的动载两类"。从理解原文的角度看，该译文传达了原意，符合"信"的原则。但由于汉语习惯用小句，这47个字的长句很难一气读完，有欠"达、雅"，如采用化整为零的分译法，则可收到醒目易读的效果。

(2) And confidence is growing in the debt-restructuring progress, the infuriatingly untidy effort that puts debtor nations on the International Monetary Fund's diet of hardnosed monetary policy, curtailed government spending, and fewer import.

译文：人们对以下诸方面的信心正在增强：对重组债务进程的信心正在增强；对改变债务国执行国际货币基金组织的严格货币紧缩政策时所持的令人气愤的拖拉作风的信心正在增强；对削减政府开支的信心正在增强；对减少进口的信心正在增强。

原文中，句子中的介词in后接四个名词短语做宾语，在第二个名词短语中含有一个定语从句。翻译时，由于四个名词短语都是由介词in统管，因此要将这四个名词短语置于同等的位置上。这种情况在汉语中往往采用平行结构，另外，在翻译介词in时，可以重复使用"对……方面"的表达式，这种重复表达在多数时候都能使译文表达效果得到强化。这里的汉语译文正是通过重复地表达"在……过程中/方面的信心正在增强"，将四层意思组织到一起。

四、变序译法

变序译法，就是把复杂的原句结构打乱，根据汉语的表达习惯重新组织句子，或者按由远及近的顺序从中间断句，层层展开，最后画龙点睛，突出主题。这种重组主要是由于原文的语序在译文中无法得到完全的保留和再现，为符合目的语的表达习惯和思维模式而在译文中对原文语序进行重新排列。

在长句翻译中，变序译法使用范围较广，"由于彻底摆脱了原文的语序和句子形式的约束，因此，比较易于做到汉语行文流畅、自然，一气呵成，有经验的译者还可以因此而使译文平添文采"（刘富庆）。例如：

(1) The reason that a neutral body is attracted by a charged body is that, although the neutral body is neutral within itself, it is not neutral with respect to the charged body, and the two bodies act as if oppositely charged when brought near each other.

译文：虽然中性物体本身是不带电的，但对于带电体来说，它并非中性；当这两个物体彼此接近时，就会产生极性相反的电荷的作用。这就是中性物体被带电体吸引的原因。

该句采用了典型的变序译法。当然，在翻译时稍不留神，译者就很容易将其译为："中性物体被带电体吸引的原因在于，虽然中性物体本身是不带电的，但对带电体来说，它不是中性，当这两个物体彼此接近时，就会产生带有相反电荷的作用"。这种误译主要是不恰当地采用了顺序分译法，以致译文内部衔接松弛，破坏了作者所提出的概念的完整性。从"although..."到句尾，都是说明"中性物体被带电体吸引的原因"，这一整体不容分割。试从"although"引导的让步状语从句入手，将原文中前置的主要信息在译文中后置，起到了画龙点睛的作用。

(2) This hope of "early discovery" of lung cancer followed by surgical cure, which currently seems to be the most effective form of therapy, is often thwarted by diverse biological behaviors in the rate and direction of the growth of cancer.

译文：人们希望"早期发现"肺癌，随之进行外科治疗，因为外科治疗可能是目前效果最好的办法。然而，由于肺癌生长速度和生长方向等生物学特征有很大差异，早期发现的希望往往落空。

翻译过程中，将主语"hope"拆分为二，先处理谓语动词之前的主语部分，包括主语和一个非限制性定语从句，然后跳到条件状语从句，最后译出主句的谓语部分。

五、综合译法

翻译中，如果顺序译法、逆序译法、分译译法、变序译法等翻译方法不能解决所有问题，就需要考虑综合使用。如果英语长句单纯采用其中任何一种方法都能译出，就需要仔细分析，或按照时间的先后，或按照逻辑顺序，顺逆结合，主次分明地对全句进行综合处理，以便把英语原文翻译成通顺忠实的汉语句子。使用综合译法，不仅可以灵活变通长句的语序，还能使译文的句法通顺自然，更符合汉语的表达习惯和中国人的思维表达方式。例如：

(1) It was wonderful to see his face shining at us out of a thin cloud of these delicate fumes, as he stirred, and mixed, and tasted, and looked as if he were making instead of punch, a fortune for his family down to the latest posterity.

译文：当他搅拌、混合、品尝时，我们看到他的脸在透着各种美妙气味的薄雾中闪着光芒,他看起来像是在为他的家族和后代创造财富,而不是一时兴起,这真是太好了。

分析原文,很容易发现句子中有四层意思:看了很不错;透过各种美妙的气味薄雾,我们可以看到他脸上闪闪发亮;在他搅和原材料、品尝味道时;看起来仿佛他是在忙着为家族和后代积累财富,而不是一时玩闹。汉语的习惯是先叙述后表态,先发生的事情先说。因此,翻译时,译者必须注意顺逆结合,交错出现。

(2) Digital computers are all similar in many ways such as having at least the five parts which all work together when solving a problem.

译文:各种数字计算机在许多方面都相似,例如,至少具备五个部分,这五个部分在解决问题时一起工作。

原文虽不是很长,但翻译起来并不容易。此句由一个主干句、一个定语从句和一个以 when 引导的状语组成,主句中含有一个以 such as 引导的同位语短语。全句有三层意思:各种数字计算机在许多方面都相似;至少具备五个部分;这五个部分在解决问题时一起工作。实际上,这三层意思都具有相对的独立性,因此,译文可以化整为零,分别叙述,分成三个单句。从内容上来看,原文逻辑关系和表达顺序与汉语基本一致,但因状语放在谓语动词之后,和汉语表达习惯不同,所以翻译时可顺中有逆,进行综合处理。

综上所述,在翻译长句时,译者的首要任务是理解原文,然后选择恰当的翻译方法重新组织出来,可以说,翻译方法选择恰当与否,会直接影响译文的好坏。

 **Exercises 2**

Translate the following paragraph into Chinese.

(1) Soil pollution is defined as the build-up in soils of persistent toxic compounds, chemicals, salts, radioactive materials, or disease causing agents, which have adverse effects on plant growth and animal health.

(2) A soil pollutant is any factor which deteriorates the quality, texture and mineral content of the soil or which disturbs the biological balance of the organisms in the soil. Pollution in soil has adverse effect on plant growth.

(3) More than 90% of hazardous waste is produced by chemical, petroleum and metal-related industries and small businesses such as dry cleaners and gas stations contribute as well.

Part 3　Writing

学术论文的撰写(七)——学术论文正文的写作(上)

如果将前言比作帽子,那么正文就是论文的躯干。下面我们分两次讲解学术论文正文部分的写作。

一、正文的结构

正文的结构包括两个方面的内容：
（1）正文是由一个或几个中间段组成的，用来讨论论文的主题；
（2）每个段是由一个或若干个句子组成的，用来叙述和讨论论文中的某一方面。一般来说，一个完整的中间段中的句子又由三个部分构成：主题句（topic sentence）；扩展句，也叫支撑句（developing sentence/supporting sentence）；结尾句或结论（conclusion sentence）。正文一般放在前言之后，空一行进行编排。

二、正文的写作方法

正文写作是科技英语写作中各种写作方法的综合运用。因为正文的内容必须将科研工作的目的、研究过程、实验方法、实验结果、数据分析及结论等完整、准确地用文字表现出来，所以论文的表达必须准确、简洁、清晰。

英语科技论文中的正文一般有两种写法，即按时间顺序和按认知逻辑顺序。

1. 按研究工作进程的时间顺序依次叙述

在研究工作实践中，研究是分不同层次进行的，经过多次循环，深化对研究内容的认识。所以，在论文正文的写作中也要按照认识问题的先后顺序来写。安排问题的序列，符合认知的逻辑性。文章开始时，先对整个工作过程的层次应略有交代，然后，把一个问题作为一个层次内容，层层有实验结果，有小结，有导出下一层次工作的引子，最终有综合，有分析，有总的观点和结论。

例如，一项研究工作的第一层次是在实验室设计实验，再现某一自然现象，使自然现象能在实验室中重复，从而抽象地推导出现该现象的条件。第二层次是研究该现象存在的原因及其本质，第三层次是研究防止出现该现象的措施，第四层次是生产中的验证和改革生产工艺。

在正文中对上述研究成果进行描述时，每一层次有明确的小结，同时提出下一层次的研究方向，这样论文的正文就会层次分明，段落清楚，下笔也就自然多了。下面是一个按时间顺序排列正文内容的实例。

【例1】

When a scientist starts out to solve a problem, he proceeds in this way. Firstly, he states his problem clearly. Secondly, he collects all the facts related to the problem that he can find. Thirdly, he keeps careful records of all the facts collected. Fourthly, he studies his facts for clues that may lead to a solution to the problem. He states the possible solution or solutions that indicate as temporary to be further tested by more observations or experiments or both. He tests each solution many times.

Finally, he arrives at a solution which agrees with all the known facts. This is the solution he states as a conclusion. Even this is not the end, for he then submits his conclusion to his fellow scientists for further testing by new experiments or new observations or both. His conclusion may then have to be modified to fit new facts.

例 1 是一个严格按时间顺序完成的段落写作。在第一段中，作者在第一句话中对科学家解决问题的方法给了一个概述，同时用 he proceeds in this way 为全段的展开进行了引导；然后作者又按科学家通常解决问题的步骤的先后顺序展开了正文段落的内容。

第一段文字中的最后一句话既是对这一段文字内容的总结，又是对下一段文字的展开进行的引导，成为下一段文字的引子。在第二段文字中，第一句话不但是第一段文字内容的继续，还是第二段文字的主题；然后第二段围绕着"已知的事实"展开了讨论。第二段文字的最后形成文字内容的循环，表示科学家解决问题完成了一个阶段。

2. 按逻辑顺序进行叙述。

它与第一种时间排列顺序法不同，它是把研究工作全过程中多次由实践到理论的循环融合起来，提炼出典型的材料和观点，按认识过程由低级向高级阶段的演变规律，逻辑地排列成章节，而不是按实验工作的原有时间顺序。一般首先介绍实验用的材料、实验设备、实验过程和实验结果。然后根据实验过程的步骤和实验结果，将数据和观察到的现象整理出来，加以综合，分别从实践上升到概念和判断。最后再进行必要的讨论，归纳出结论，完成推理。按逻辑排列的正文的写作方法有两种，一种是划分法，一种是分析法，分别叙述如下：

（1）划分法

划分法是把事物划分为几个组成部分，分别予以处理。如一篇文章讲的是防止土壤侵蚀问题，可以根据内容将其分为四点，如：

Generalization（概述）
item Ⅰ　item Ⅱ　item Ⅲ　item Ⅳ

Ⅰ. Erosion control in the East

Ⅱ. Erosion control in the South

Ⅲ. Erosion control in the North

Ⅳ. Erosion control in the West

然后再根据每一点的内容再细分，如：

Ⅰ. Erosion control in the East:

a. Erosion control in the past

b. Erosion control in the present

c. Erosion control in the future

由于这种划分非常方便，并且有条理、有层次，十分合乎逻辑，所以在科技论文的正文写作中常常使用这种方法。下面就是一个按照逻辑排列顺序的实例。

【例 2】

The amount of energy released from one atomic bomb is far greater than can be released at one time from ordinary sources of energy. This is because the atomic nucleus itself is burnt apart, and the nucleus of an atom has the most concentrated store of energy known. What is called atomic energy is actually nuclear energy.

There are two known ways to release the energy from the nucleus of an atom. One way is to split a nucleus of an atom into other nuclei of smaller mass. This method is called fission. The particles of the split atom produce heat through friction as they fly through the air. In addition, the reaction produces gamma rays and other tiny particles—neurons—that can start fission in nearby atoms. The release of energy from a chain of these fission reactions in a pound of Uranium-235 is greater than the release of energy from burning 2,600,000 pounds of coal.

The second known way to obtain energy from the nucleus is to bind nuclei together to form heavier nuclei. Some of the mass of the combining nuclei becomes energy. This process is called fusion. It is the sourer of the tremendous energy of the sun.

Much more energy can become available as the result of fusion than as the result of fission. Fusion is the principle of the hydrogen bomb. Scientists are now trying ways to control the tremendous energy released by fusion.

例2的第一段是概括性引言，由于它的最后一句话是中间段落的引子，所以也将其摘录。我们可以把第二段、第三段和第四段的文章内容按下述要点进行。

Ⅰ. Atomic energy is actually nuclear energy.

Ⅱ. Fission

a. splitting a nucleus of an atom

b. production of heat

c. the results of fission

Ⅲ. Binding nuclei together to form heavier nuclei

a. production of energy—fusion

b. an example—source of the sun's energy

Ⅳ. Conclusion

a. more energy from fusion

b. the principle of fusion

c. control fusion energy

同时，例2清楚地显示第一段和第三段的最后一句话都是为下一段的展开进行必要的引导，从而使整篇文章之间的段落衔接更加紧密。

（2）分析法

分析法是由作者提出需要解决的问题，需要采取的措施或需要解释的现象，然后对其进行分析，直接找到解决这些问题的方法，或说明所采取的针对性措施的好处与不足，或解释清楚造成某些现象的原因；也可对几种可供选择的方法进行解释和对比，并证明为什么某一种选择是最优的。

分析法常常用下面的方式表示：

Item Ⅰ

Item Ⅱ

Item Ⅲ

Item Ⅳ

Generalization

分析法实际上是归纳写作手法的体现。但因其对写作人员的文字功底要求较高,通常情况下若不是进行多项情况的比较,一般不采用。

三、主题句(topic sentence)的写作方法

主题句不仅表达了一个段落的中心思想,也指出了此段内容的发展方向与方式。主题句通常位于一段文字的句首,但也可放在中间或结尾。主题句的位置不同有着不同的作用。句首是主题句最常见的位置,这样可以让读者较容易地了解整段的内容。放在中间是因为这一段的前一部分内容要作为本段所叙述内容的背景或需要连接前段内容。放在最后则是采用了归纳法的写作方法,将该段文字的中心内容放在本段结尾,予以总结。通常,科技英语的段落写作往往把主题句放在段首,然后通过扩展细节和提供依据对主题句的内容进行论述。

下面是几个主题句的使用方法的实例。

【例3】

Topic Sentence: Jefferson was a good and tireless writer.

Evidence: 1. fifty volumes works

2. author of Declaration of Independence

3. effect of his writing

Paragraph:

Jefferson was a good and tireless writer. His complete works, now being published for the first time, will fill more than fifty volumes. His talent as an author was soon discovered, and when the time came to write the Declaration of Independence at Philadelphia in 1776, the task of writing it was his. Millions have thrilled to his words, "We hold these truths to be self-evident, that all men are created equal…"

例3中,主题句是Jefferson was a good and tireless writer。围绕着good和tireless,作者使用了供扩展这一中心内容的支撑文字内容。把good和tireless用"50卷书、独立宣言的作者"以及"他的作品所带来的影响"这些事实进行了翔实的说明,完整地体现了主题句的内容。

【例4】

Topic Sentence: But I think it is a pest and a time waster.

Evidence: 1. impossible to escape from idle or inquisitive chatter

2. it rings when you don't want it to

Paragraph:

Many people think a telephone is essential. But I think it is a pest and a time waster. Very often you find it impossible to escape from idle on inquisitive chatter-box, or from somebody who wants something for nothing. If you have a telephone in your own house, you will admit that it tends to ring when you least want it to ring: when you are asleep, or in the middle of a meal or a conversation, or when you are just going out, or when you are in your bath. Are you strong-minded enough to ignore it, to say to yourself, "Ah, well, it will all be the same in a hundred years' time?" You are not. You think there may be some important news or message for you. I can assure you that if a message is really important it will reach you sooner or later. Have you never rushed dripping from the bath, or chewing from the table, or dazed from the bed, only to be told that he or she is dialing wrong number?

例4中，主题句是在第二句。作者在整段文字中围绕着 pest and a time waster 进行了叙述。

【例5】

Topic Sentence: It just happens that Americans and Japanese have a different way of looking at human relationships and thus have a different way of showing respect.

Evidence: 1. Americans might be embarrassed on a formal occasion

2. Japanese might feel insulted on a casual occasion

3. forms of greeting

Paragraph:

Americans might be embarrassed because their Japanese friends are so formal with them. Japanese might feel insulted because American acquaintances greet them casually. Still, the forms of greeting in both countries only show respect for others. It just happens that Americans and Japanese have a different way of looking at human relationships and thus have a different way of showing respect.

例5的主题句是段落最后一句话，整段文字所叙述的内容是围绕着主题句展开的。主题句放在段尾是采用了归纳法的写作技巧。

综上所述，主题句是文章段落的中心思想的体现，无论其位置如何，都应体现这一原则。

需要指出的是，有人曾对英美论文做过统计，发现60%的主题句在段首，30%的主题句在段尾，10%的主题句在段落的中间。对于初学者来说，最好将主题句放在段首，以便使段落的内容按顺序展开。

下面就如何写好主题句和主题句写作时应注意的事项进行叙述。

1. 主题句的结构

要写好主题句，首先要弄清主题句的结构。主题句的结构通常包括两个部分，即中心议题（topic/ subject part）和控制部分（controlling idea）。

中心议题就是该段文字所要涉及的人、事或问题，而且整个段落的文字内容都要围绕着这个人、这件事或这个问题展开。

例如，在前述例3中主题句的中心议题是Jefferson；例4中，主题句的中心议题是it（指telephone）；例5中主题句的中心议题是Americans和Japanese。因此，主题句的中心议题是文章所要叙述或讨论的对象。

在主题句中，还有一个部分是控制部分。控制部分是决定该段文字的发展方向或扩展方法，它可能是一个词、短语或从句。控制部分暗示这段要由时序法（列举法）、举例法、归纳法、对比法或因果法等中的一种或几种方法混合在一起来发展。例如，有关篮球（basketball）这个主题，就可以写出下面5个主题句，每个主题句的控制部分都不一样，而且所使用的写作技巧也不一样（见画线部分）。

【例6】

1. Basketball has become more popular *within the past twenty years*. （时间）
2. Basketball and volleyball *have a great deal in common*. （归纳）
3. Basketball is *less dangerous than football*. （对比）
4. Basketball is popular *for several reasons*. （原因）

例6的4个句子中，每个例句中的控制部分所表达的内容不一样，因此在各自例句后的段落中的文字内容自然也就不一样。第1句的文字内容是叙述在近20年中篮球运动的普及情况。第2句则是讲述篮球与排球之间的共同之处，其后的文字就要说清这两项球类运动之间的共性。第3句的控制部分是将篮球与足球运动之间的危险性进行对比，那么该段的文字内容也必定要沿着这个方向发展。第4句控制在几个原因上，作者要在本段的文字中对这几个原因阐述清楚。

综上所述，主题句的构成是由中心议题与控制部分组成，中心议题是所讨论的对象，控制部分是关于所讨论的对象的具体内容，同时也限定了文字论述时所要运用的技巧。

2. 主题句的写作技巧与文体要求

主题句的写作质量如何会直接影响到整篇文章的文字质量。因此，主题句写作的质量对一篇科技英语文章的成功与否起着至关重要的作用。

怎样写好主题句？在写好主题句时要注意哪些问题？主题句的文体要求又有什么特点？这些都是需要解决和注意的问题。

（1）主题句必须有该段要讨论的中心议题（topic）和控制该段发展的范围（controlling idea），而控制范围常常用一个或几个关键词来表达。

【例7】

1. These experiments have been done in the newly-built laboratory.
2. There exists air pollution in China.
3. Noise pollution is becoming more and more serious in this country.

例7中第1句没有中心议题也不是该段的发展范围，无法继续发展下去。第2句虽有中心议题（air pollution），但没有关键词来控制，使读者不知下一步要涉及的范围。第3句既有中心议题（noise pollution）又通过关键词 more and more serous 来控制该段的内容，因此，第3句应该是一个较好的主题句。

（2）主题句不能过于宽泛或概括化，这是由主题句的文体要求与写作特点所决定的。因为主题句太宽泛，整个段落就会显得太松散，从而对文章的内容难以控制。有的主题句太宽泛，给人的感觉就是这不是一个段落的主题句，而是一篇文章的题目或是一本书的书名。

【例 8】 Health and wealth, which is better?

例 8 中的主题句作为一篇文章的题目是合适的，而作为一个段落的主题句则显得太宽泛、太笼统。现将其作为一篇文章的题目，然后再根据其内容扩展为几个主题句，将题目细化。

【例 9】

1. Some people hold that wealth is more important than health.

2. However, the majority of people believe that health is more important than wealth.

3. In my opinion, health is better than wealth.

在例 9 的主题句中，每个句子都可作为一段文字的主题内容，在第 1 句中主题所控制的扩展方向是 wealth is more important than health，而且还限定了 some people 这个范畴。在第 2 句中，表示了相反的观点，即 health is more important than wealth，它所要求的扩展内容也只能按这个方向进行。第 3 句则是作为个人的观点提出一个结论性意见，其限制的内容就是：health is better than wealth。在例 9 的三个主题句中，语言结构简单，文字表达内容准确、具体，控制部分要求明确，达到了主题句的写作要求。

（3）主题句写作也不能过于具体、确切。因为太具体、太确切的主题句难以在随后的段落文字中对其进行扩展，即它所叙述的是个具体、翔实的内容，再也没有进一步发展与说明的必要。例如下列一组例句：

【例 10】

1. If there were no electric power, people would have to use fire to light the room in the evening. （太窄，无法展开）

2. If there were no electric power, things would be quite different. （清楚，明确，恰当）

例 10 中的第 1 句已经道出了事实，而且还提供了具体的细节，应该说对其是无法进行扩展的。而第 2 句的主题句使用了 things would be quite different 这个发展方向，使作者可使用列举的写作方法对 quite different 进行必要和具体的描述。

（4）因为主题句要清楚、直接地表达一个中间段的中心思想，所以主题句的句型常常是简单句或简洁的句子。试比较下列一组实例：

【例 11】

1. In the United States, *the system* of forced labor, which was known as slavery, lasted almost 250 years. （冗长）

2. *Slavery* lasted almost 250 years in the United States. （简明，清楚）

例 11 中，第 1 句话使用了非限制性定语从句，还有一些不必要的重复。而第 2 句只使用了一个简单句，句中首先点明了 slavery 这个主题，然后用时间来表明下述文字中应扩展的方向，文字表述得准确、清楚、简明，是一个质量较高的主题句。

 **Exercises 3**

Have you written something about your daily discharge of MSW? Please write few topic sentences with topics and controlling idea, then to expand those sentences into paragraphs.

 Expanding Reading

The UK Soil and Herbage Survey (UKSHS)

The UK Soil and Herbage Survey (UKSHS) is a comprehensive survey of the concentrations of major contaminants in soils and herbage across the UK. It provides a reliable baseline against which intensive local surveys and future national surveys can be assessed. For some contaminants, the UKSHS provides us with the first comprehensive picture of their concentrations across the UK.

The UKSHS determined the concentrations of 12 metals and arsenic, 22 polycyclic aromatic hydrocarbons (PAHs), 26 polychlorinated biphenyls (PCBs) and 17 polychlorinated dioxins and furans (dioxins) in soil and herbage at 122 rural, 28 urban and 50 industrial locations.

The results in UKSHS show that, for all the metals studied (and arsenic), concentrations in industrial soils are significantly higher than in rural areas. Soil concentrations of copper, lead, mercury, nickel, tin and zinc are higher in urban sites compared with rural sites.

There are differences across the four countries of the UKSHS, partly reflecting geology but partly the result of differing anthropogenic inputs. Concentrations of titanium in rural and urban soils from Northern Ireland and Scotland are significantly higher than in England, reflecting the occurrence of titanium-rich basaltic rocks in these areas. Nickel and chromium are also higher in urban soils from Northern Ireland. The results from the UKSHS were compared with those from earlier work to estimate trends in metal concentrations over time. The inevitable differences in methodology mean that such trends are, at best, approximate, but the data suggest that cadmium and copper levels in soil increased between the mid-1800s and the 1980s. For lead, nickel and zinc, the trends are inconclusive.

The UKSHS results for PCBs indicate that, despite restrictions on their production since the mid-1970s, urban and industrial areas are still significant sources of PCBs to the environment—though the concentrations of PCBs in soil have fallen approximately 800-fold over that period. Concentrations of total PCBs in urban and industrial soils and herbage are approximately 1.5 times those in rural areas. There are differences in PCB concentrations in rural soils and herbage across the four countries of the UK, but the patterns are complex and the data strongly skewed. PCB concentrations in rural soils are highest in Scotland and lowest in Northern Ireland, with England and Wales intermediate. Puzzlingly, PCB concentrations in rural herbage are highest in Northern Ireland despite it having the lowest soil concentrations. Concentrations of total PCBs in urban soils in Scotland, Wales and Northern Ireland are significantly lower than in England. The findings on both total PCB concentrations and the pattern of individual PCBs are not consistent with high temperature sources or spillages being significant contributors of PCBs to the UK environment.

(*https://assets.publishing.service.gov.uk/government/uploads/system/uploads/attachment_data/file/ 291146/scho0307bmep-e-e.pdf*)

UNIT 5 Environmental Management and Environmental Impact Assessment

Lesson 1　Administration on Environmental Management

Part 1　Reading

1　Introduction

China's current environmental protection management system was established in accordance with the Constitution of the People's Republic of China (the Constitution) and the Law of the People's Republic of China on Environmental Protection (EP Law)[①]. At the central government level, the State Council plays a leading role in directing and administering environmental protection programs, while the Ministry of Ecology and Environment (the MEE) carries out the actual supervision and administration of the national environmental protection programs. Other ministries and administrations also play significant roles in regulating the environmental affairs within their scope of responsibilities. Finally, local governments at all levels are responsible for environmental issues within their respective jurisdictions. Various administrative departments play different roles in environmental supervision and management.

2　The History of the MEE

The construction and development of environmental protection agencies in China was initiated in the 1970s. In 1971, the Leading Group of The Utilizations of Three Wastes (exhaust gas, wastewater and solid waste) was established by the National Planning Committee. In 1973, the State Council held the First National Conference on Environmental Protection, in which the setting up of an administration and management agency for environmental protection was proposed. In October 1974, the Leading Group of Environmental Protection (the LGEP) was formally founded under the auspices of the State Council to mark the commencement of the formation of China's environmental protection agencies. The main responsibilities of the LGEP were to institute policies, principles, and guidelines for environmental protection; to review and approve national environmental protection plans; and to organize, coordinate, supervise and inspect the implementation of environmental protection within relevant departments and at all levels[2]. Directly under the LGEP, an executive Environmental Protection Office was established and was to be responsible for routine work. In 1978, various environmental protection institutions, namely, environmental protection bureaus (EPBs) or environmental protection offices (EPOs), were launched in municipalities directly under the Central Government, autonomous regions, and industrial cities.

In 1982, following a decision of the 23rd Meeting of the Standing Committee of the 5th National People's Congress (NPC), the Ministry of Urban and Rural Construction and Environmental Protection (MURCEP) was founded by incorporating the State Construction Commission, the State Administration of Urban Construction, the State Administration of Construction Projects, the National Administration of Surveying and Mapping, and the LGEP.

In May 1984, the Environmental Protection Commission (EPC) was established directly under the State Council to deliberate and finalize relevant policies and guidelines for environmental protection; to formulate environmental plans; and to lead, organize, and coordinate national environmental protection efforts. One of the Vice Premiers of the State Council also functioned as the Chairman of the Environmental Protection Commission, which operated through its office under the MURCEP. In December 1984, the Environmental Protection Bureau under the MURCEP was reshuffled into the National Environmental Protection Administration (NEPA), which was still under the leadership of both the MURCEP and the Environmental Protection Commission of the State Council, to be mainly responsible for the planning, coordination, supervision, and guidance of national environmental protection efforts[3].

In July 1988, the responsibilities for environmental protection were transferred from the MURCEP to the NEPA, which was then defined as an affiliated institution of the State Council (sub-ministerial level) and the office of the Environmental Protection Commission of the State Council to function as a competent department under the State Council for integrated environmental management. In June 1998, the NEPA was upgraded to the State Environmental Protection Administration (SEPA; ministerial level), as an affiliated institution of the State Council taking charge of the environmental protection effort. The Environmental Protection Commission of the State Council was revoked, and in July 2008, the SEPA was elevated to the Ministry of Environmental Protection (MEP). As an integral department of the State Council, the MEP is the central competent authority for environmental protection and management, with its major functions being the establishment of an integrated environmental protection system and the management and monitoring of environmental pollution prevention and control. In March 2018, the MEP was then reorganized as the Ministry of Ecology and Environment (MEE) as a consequence of the institutional reform of the State Council.

3　Structure and Functions of the MEE

The MEE is located in Beijing and is made up of a number of functional departments and institutes, together with 12 dispatched agencies located in various regions. There are 23 affiliated public institutions (including scientific research institutes, publishing agencies, monitoring departments, and so on) and five social groups (including academic associations, environmental associations, foundations, etc.) under the direct command of the MEE. The key functions of the MEE can be summarized as follows: formulation of environmental protection regulations, enactment and supervision of environmental planning for key regions and river basins, adjudication of major environmental pollution accidents and ecological damage incidents, and management of other administrative activities (Figure 5.1).

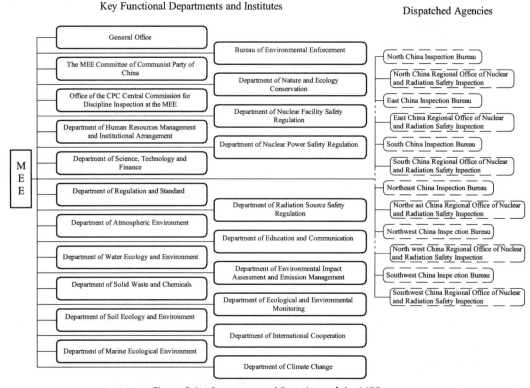

Figure 5.1　Structune and Functions of the MEE

 **Words and Expressions**

Constitution of the People's Republic of China	*n.* 中华人民共和国宪法
the Law of the People's Republic of China on Environmental Protection	*n.* 中华人民共和国环境保护法
the State Council	*n.* 国务院

the Ministry of Ecology and Environment (the MEE)		n. 生态环境部
administration	[ədˌmɪnɪˈstreɪʃn]	n. 管理，行政，施行，执行，（企业、学校等的）管理部门
jurisdiction	[ˌdʒʊərɪsˈdɪkʃn]	n. 司法权，审判权，管辖权，管辖范围
the Leading Group of The Utilizations of Three Wastes		n. 三废利用领导小组
the National Planning Committee		n. 国家计委
auspices	[ˈɔːspɪsɪz]	n. 赞助，主办，预兆（auspice 的复数）
coordinate	[kəʊˈɔːdɪneɪt]	v. 使协调，使相配合，协同动作
implementation	[ɪmplɪmenˈteɪʃən]	n. 实施，执行，贯彻，生效
the National People's Congress		n. 全国人民代表大会
the Ministry of Urban and Rural Construction and Environmental Protection （MURCEP）		n. 城乡建设和环境保护部
the State Construction Commission		n. 国家建设委员会
the State Administration of Urban Construction		n. 国家城市建设管理局
the State Administration of Construction Projects		n. 国家建筑工程管理局
the National Administration of Surveying and Mapping		n. 国家测绘局
deliberate	[dɪˈlɪbərə]	v. 仔细考虑，深思熟虑，反复思考
the Vice Premier		副总理
the Chairman of the Environmental Protection Commission		环境保护委员会主席
revoke	[rɪˈvəʊk]	v. 撤销，取消，废除，使无效
integral	[ˈɪntɪɡrəl]	adj. 完整的，不可或缺的，必需的
the institutional reform		n. 体制改革
dispatch	[dɪˈspætʃ]	v. 派遣，调遣，派出，迅速处理
affiliate	[əˈfɪlieɪt]	v. 使附属，使隶属，加入
adjudication	[ədʒuːdɪˈkeɪʃn]	n. 判决，裁决，审判，宣告

Notes

① China's current environmental protection management system was established in accordance with the Constitution of the People's Republic of China (the Constitution) and the Law of the People's Republic of China on Environmental Protection (EP Law).

参考译文：中国现行的环境保护管理体系是根据《中华人民共和国宪法》(以下简称《宪法》)和《中华人民共和国环境保护法》(以下简称《环境保护法》)建立的。

② The main responsibilities of the LGEP were to institute policies, principles, and guidelines for environmental protection; to review and approve national environmental protection plans; and to organize, coordinate, supervise and inspect the implementation of environmental protection within relevant departments and at all levels.

参考译文：环境保护领导小组的主要职责是制定环境保护政策、原则和指南；审议批准国家环境保护规划；组织、协调、监督、检查有关部门和各级环境保护工作的落实情况。

③ In December 1984, the Environmental Protection Bureau under the MURCEP was reshuffled into the National Environmental Protection Administration (NEPA), which was still under the leadership of both the MURCEP and the Environmental Protection Commission of the State Council, to be mainly responsible for the planning, coordination, supervision, and guidance of national environmental protection efforts.

参考译文：1984年12月，城乡建设和环境保护部下属的环境保护局改组为国家环境保护局（NEPA），该局仍由城乡建设和环境保护部及国务院环境保护委员会领导，主要负责规划、协调、监督，指导国家环境保护工作。

Exercises 1

1. According to the reading material, chose the best answer(s) from the options.

(1) China's current environmental protection management system was established in accordance with _____ and _____.

A. the Constitution of the People's Republic of China

B. the Law of the People's Republic of China on Environmental Protection

C. the National Environmental Protection Administration

D. none of them

(2) At the central government level, _____ plays a leading role in directing and administering environmental protection programs.

A. the Constitution   B. the Ministry of Ecology and Environment

C. Other ministries and administrations   D. the State Council

(3) The construction and development of environmental protection agencies in China was initiated in the _____.

A. 1970s   B. 1980s   C. 1990s   D. 21st century

(4) In March 2018, the MEP was then reorganized as _____ as a consequence of the institutional reform of the State Council.

A. the National People's Congress

B. the Ministry of Ecology and Environment

C. the Ministry of Urban and Rural Construction and Environmental Protection

D. the State Construction Commission

(5) The MEE is located in _____ and is made up of a number of functional departments and institutes, together with 12 dispatched agencies located in various regions.

　　A. Chengdu　　　　　B. Shanghai　　　　　C. Beijing　　　　　D. Hongkong

(6) There are 23 affiliated public institutions— including _____ , _____ , and _____ , and so on —and 5 social groups under the direct command of the MEE.

　　A. scientific research institutes　　　　B. academic associations
　　C. monitoring departments　　　　　　D. publishing agencies

2. Translate the first paragraphs into Chinese.

Part 2　Translation

英语翻译中从句的翻译（一）——主语、宾语、表语、同位语从句

科技英语中有大量的长句，除了谓语之外的每一种句子成分都可以由从句充当。英语的从句主要包括名词性从句、定语从句和状语从句等几种类型，它们在功能上都是用来修饰、叙述或补充说明主句或句中的某一短语。我们首先来介绍一下名词性从句。

英语名词性从句包括若干种从句，如主语从句、宾语从句、表语从句、同位语从句等。这些从句在句中充当主语、宾语、表语或同位语等成分，其功能相当于名词、名词性词组或短语，因此得名名词性从句。英语中名词性从句着实不少，也主要是由关系代词（what, whatever, whoever）、关系连词（that, which）和关系副词（where, why, when）等引导。

一、主语从句

英语中的主语从句主要有两种：一种是主语从句位于主句之前；一种是主语从句位于主句之后。由从属连词、连接副词或连接代词引导的句子一般位于主句之前，而由先行词 it 引导的特殊句子位于主句之后。对于这两种句型，因具体情况不同，翻译时也需有所不同。一般来说，可采用顺序译法、逆序译法、综合译法和分译译法。例如：

(1) Whether that UFO was a spaceship from outer space or just a flock of flying birds still remains a puzzle.

译文：不明飞行物是来自外太空的太空船，还是仅仅是一群飞行的鸟，仍然是个谜。（顺序译法）

(2) If balance sheet were made, it would be found that the oil ban has created more difficulties for the European countries than for OPEC.

译文：要是编出一份决算表，就会发现石油禁运给欧洲国家造成的困难要比给石油输出

国造成的困难大得多。（顺序译法）

(3) It seemed inconceivable that the pilot could have survived the crash.

译文：驾驶员在飞机坠毁之后，竟然还能活着，这真是难以想象的事。（逆序译法）

(4) It seems possible that the earth might not be the unique planet with life.

译文：地球似乎不可能是唯一有生命的星球。（综合译法）

(5) It should be realized that magnetic forces and electric forces are not the same.

译文：应该认识到，磁力和电力是不相同的。（分译译法）

二、宾语从句

一般来说，宾语从句包括动词宾语从句和介词宾语从句两种。

1. 动词宾语从句

动词宾语从句的翻译方法与主语从句的翻译方法大体相同，可采用顺序译法、逆序译法和综合译法三种翻译方法，例如：

(1) Smelting experiments in this furnace demonstrated that is was possible to carry out complete desulphurization of sulphide concentrates autogenously.

译文：在该炉上进行的冶炼试验表明，硫化精矿完全脱硫是可行的。（顺序译法）

(2) But at that time no one could explain why the path of a planet must be an ellipse.

译文：但是，为什么行星的轨道必须是椭圆的，当时谁也说不清。（逆序译法）

(3) He asked only that he be allowed to continue his work undisturbed.

译文：他唯一的要求是允许他不受干扰地继续工作。（综合译法）

2. 介词宾语从句

在介词后做宾语的句子叫介词宾语从句。这种从句往往用连接代词 what which/who 或连接副词 how/when/why/where/whether 等引导。当然，有时也可以和连词 that 连用，但必须是用于介词 in/except 之后。一般可采用顺序译法、逆序译法、综合译法、分译译法、转换译法。例如：

(1) As a result of what the astronauts have done, the world has never been closer together before.

译文：由于宇航员所作的努力，世界从来没有像现在这样紧密相连。（顺序译法）

(2) There seems to be no end to what petroleum can do for man.

译文：看来石油可造福于人类的用途将是无穷无尽的。（逆序译法）

(3) The question then arises of how this inertial mass compares with the gravitational mass.

译文：然后就出现了如何比较惯性质量和引力质量的问题。（综合译法）

(4) Plastics are different from other materials in that they possess a combination of properties.

译文：塑料与其他材料不同，因为塑料具有综合性能。（转译为原因状语从句）

三、表语从句

表语从句较之其他从句更好判断和识别，它总是位于系动词之后，对其主语起到解释的作用，语序和汉语基本一致。表语从句一般用顺序译法和逆序译法。例如：

(1) One of the important properties of copper is that it conducts electricity better than other materials.

译文：铜的重要特性之一是其导电性能好。（顺序译法）

(2) One of the important properties of plastic is that it does not rust at all

译文：不会生锈是塑料的重要特性之一。（逆序译法）

四、同位语从句

名词性从句在句子中可以做同位语，简称为同位语从句，用于进一步说明前面某一名词的内容。同位语从句与定语从句容易混淆。首先，定语从句中的 that 是关系代词，在从句中充当主语或宾语等成分；而同位语从句中的 that 则不然，它是连词，在句子中只是起到连接作用而不具有任何实质性的意义。其次，定语从句对其先行词只起修饰和限定作用，而同位语从句则在内容上起到更进一步解释和说明先行词的作用。由于同位语从句和定语从句具有一定的共同点，所以在翻译处理的方法上也大同小异，可采用顺序译法、逆序译法、综合译法等。例如：

(1) But his findings gave some support to the idea that fusion may be possible without extreme heat.

译文：可是，他的发现却支持了这样的观点：聚变可以在没有极高温的情况下产生。（顺序译法）

(2) There can be no question about the fact that industrialization does raise living standards.

译文：工业化确能提高生活水平，这是毫无疑问的。（逆序译法）

(3) Even the most precisely conducted experiments offer no hope that the results can be obtained without any error.

译文：即便是最精确的实验也没有希望获得毫无误差的实验结果。（综合译法）

(4) The theory that diseases are caused by bacteria was advanced by Pasteur, a French chemist.

译文：细菌致病的理论是法国化学家巴斯德提出来的。（同位语转译为定语）

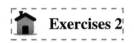 **Exercises 2**

Translate the following paragraph into Chinese.

(1) Stripping electrodes thus represent a unique type of chemical where the recognition (accumulation) and transduction (stripping) processes are temporally resolved.

(2) The remarkable sensitivity of stripping analysis is attributed to its preconcentration step, in

which trace metals are accumulated onto the working electrode.

(3) Several electrochemical devices, such as pH or oxygen electrodes, have been used routinely for years in environmental analysis.

(4) The reason that a neutral body is attracted by a charged body is that, although the neutral body is neutral within itself, it is not neutral with respect to the charged body, and the two bodies act as if oppositely charged when brought near each other.

Part 3　Writing

学术论文的撰写（八）——学术论文正文的写作（下）及结尾的写作

一、扩展句（supporting/developing sentences）的写作方法

在完成一个段落的主题句后，就应按照主题句中的控制部分来扩展该段落的内容，以完成整个段落的写作。具体地说，要对控制部分的内容进行讨论、说明、支撑或者证明等。那么，在扩展句中所要使用的是事实、理由、原因、说明、图表、比较、对比、定义等方式来阐述主题句中控制部分的思想内容。所以，一段写作质量较高的文字必须遵循以下三个原则：一致性、连贯性和完整性。

1. 一致性（unity）

一段好的文字最重要的特点是一致性。在写作一个段落时，只能讨论问题的一个方面或一个内容，这就叫文章的一致性。所以，在一段文字中，所使用的事例、所叙述的事实以及所解释的理由，都必须与所叙述的主题有关。

【例1】

Japanese women have changed since the war. They have become prettier brighter, more decisive, and more outspoken. The young women certainly are far more logical and far less sentimental than the prewar generations. Some regret this. They think women, in gaining their freedom, have lost. Their feminity—their modesty, their warmth, their shy grace. They accuse women of being drawn to superficial things. A modern Japanese woman, they say, instead of trying to enrich her inner self, is in a mad scramble to ape anything that is new and foreign—fashions, cosmetics, hairdos, rock-and-roll. And there are many Japanese who say that a caricature of an up-to-date wife is one who sits beside a washing machine in a house that has no hot running water.

例1的主题是第1句，围绕日本妇女在第二次世界大战后所发生的变化而展开。主题句之后的扩展句就是围绕着change展开的。作者运用对比、比较、陈述等手法叙述了日本妇女自第二次世界大战以来的变化情况。由于所述文字内容与主题句的要求是一致的，应该说例1是一段写得成功的文字。

2. 连贯性（coherence）

连贯性就是要在一段文字中将所有的句子清楚地、有逻辑地联系在一起，共同说明这一段的主题。在表达连贯性很强的段落中，每一句话都应自然地从前一句中繁衍而出，对该段的中心思想进行扩展。理想地说，这样的一段文字应具有一种自然流畅感，读者也就轻而易举地发现所有这一段中的句子从逻辑上是联系在一起的。所以在这样的一段文字中，写作手法可采用时间顺序法、空间描述法、比较对比法以及按重要性来排列各句的顺序等写作手法。不论使用何种方法，首先要考虑到主题句中控制部分的要求，然后再使用不同类型的连词和代词，或重复某些关键的词汇和词组，或使用同义词或词组来表示重复。这样做既能使读者记住前面的内容，又把各句都联系起来。

【例 2】

Man has learned to travel faster and faster throughout history. *When* the wheel was invented *over a thousand years ago*, man learned that it was possible to travel faster on wheels than on foot. With the invention of the steam engine about *two hundred years ago*, man began to travel at what was called "dangerous" speed between 20 to 30 miles an hour. The gasoline engines that were used between *1900 and 1920* developed speeds up to 60 miles an hour. *In the late 1920s*, propeller airplanes began to fly at speeds of more than 100 miles an hour. *About twenty years ago*, man began to travel in commercial jet planes at speeds above 500 miles an hour.

例 2 中，所有的句子都是按照连贯性的要求，从内容和形式两个方面的原则进行了自然流畅的连接。从画线部分可以看出，这一段文字的连接方式是围绕主题句 Man has learned to travel faster and faster throughout history，其中 throughout history 就是控制部分，因此，扩展句要按时间顺序来安排。时间顺序的自然推移突出了连贯性，体现出了文笔流畅、条理清楚的基本特点。

3. 完整性（completeness）

如果一个段落没有完整充分地说明主题句所要求叙述的内容，那么这个段落就不是一个意义完整的段落。完整性就是要将主题句中要求表示的内容用一致性和连贯性的方式完整地表达清楚，而不是用字数多长来进行衡量。如果主题句的限制较多，所写的段落也就应较短；反之亦然。所以，完整性应首先按照主题句所要求表示的内容提供具体、翔实而具有说服力的事实，并在主题句控制的扩展方向上进行组织。其次就是要运用好各种写作技巧，使文章段落结构严整。下面是两个围绕一个主题进行讨论的实例，试比较哪一个体现出较好的完整性。

【例 3】

At the turn of the last century, many diseases shortened human life. People did not live very long; what life they had was miserable. If disease did not kill them, poor hygiene did. Most people died from causes that no longer kill. Many were unhappy from the premature mortality around them. Families could not rely on younger members to outlast their parents. However, through improvements in medicine and public hygiene, we now live many years longer.

【例4】

At the turn of last century, infectious diseases were the primary health threat to this nation. Acute respiratory conditions such as pneumonia and influenza were the major killers. Tuberculosis, too, drained the nations vitality. Gastrointestinal infections destroyed the child population. A great era of environmental control helped change all this. Water and milk supplies were made safe. Engineers constructed systems to handle and treat dangerous human wastes and to render them safe. Food sanitation and personal hygiene became a way of life. Continual labors of public health workers diminished death rates of mothers and infants. Countless children were vaccinated. Tuberculosis was brought under control. True, new environmental hazards replaced the old. But people survived to suffer them. In 1900, the average person in the United States rarely eked out fifty years of life. Some twenty years have since been added to this life expectancy.

例3和例4可以看出，讨论的主题都是疾病对人的寿命的影响。在例3中可以清楚地看出所讨论的内容只不过是对主题句的重复；而且在主题句中对段落主要部分没有明显的控制。从第二句开始到段落结束给人的感觉是几乎每个句子都可拉出来单独作为一段文字的主题句，所以说例3中的内容缺乏一致性，而且文字中间没有任何联系，也缺乏连贯性和完整性。

例4中，作者在主题句中就明确地控制了该段文字扩展的方向，然后在扩展句中对为什么人的寿命得以延长进行了详尽的叙述，所扩展的范围包括饮水、牛奶、食品、工程技术、公共卫生等。在段落的最后一句实际是对全文一个概括性的总结，整个段落所描述的内容完整地满足了主题句中提出的内容要求。

在科技英语写作中，如果说主题句是一个人的头部，那么扩展句就应该是一个人的躯干。扩展句是对主题句中的主题部分按控制部分的要求，将其清楚、完整、准确地描述出来。因此，在写扩展句时，除了要注意一致性、连贯性和完整性外，在写作文体和其他方面还要注意以下特点与要求。

（1）文字的逻辑性

在学术写作方面，除了一致性、连贯性和完整性外，还要注意文字的逻辑性。要做到逻辑清晰，就要使用好转接词。因为英语中的转接词用来表明文章的发展线索，它们可能是连词、代词或是被重复的单词和词组。

（2）避免重复

在英语文章中，无论是科技文章还是文学作品，都要通过使用同义词、同义词词组或者代词来取代或指代前文中可以取代或指代的词汇与词组。

（3）尽量不使用意义模棱两可的修饰语

科技英语文章文体特征之一就是准确。如使用诸如 almost/nearly/about 等词汇就达不到准确表达一个概念、说明一个过程、描述一个物体的基本要求。

【例5】

① I have a certain way to make carbon dioxide turn lime water milky. （模糊）

② I have a sure way to make carbon dioxide turn lime water milky. （清楚）

③ I have some way to make carbon dioxide turn lime water milky. （清楚）

例 5 的三个句子中，关键词是做定语的 certain, some, a sure。第 1 句中的 certain 是个多义词，它可以解释为 "可靠的（sure）"，也可解释为 "某一、某些"。如果作者的意图是要明确表示，就可以用第 2 句中的 "sure"，那么该句的语意就为 "我有个可靠的方法用二氧化碳把石灰水变成乳白色"。如果作者是想说明某一种方法，那么就可以用第 3 句中的 "some"，该句的语意就为 "我有一种方法用二氧化碳把石灰水变成乳白色"。

从上述三个例句可以得出如下结论：在科技英语写作中，对容易造成歧义的词，选用时要慎重。

（4）正确使用人称

科技论文常用的人称主要体现在使用人称代词。在人称代词的使用上有两种主张，一种是传统式的主张，认为科技文章应侧重叙事和推理，读者重视的是论文的内容和观点，感兴趣的是作者的发现，不是作者本人。因此要避免使用第一、第二人称代词尤其不能使用第一人称单数 I。因为 I 是主观的象征，与科学是不相称的，所以导致了被动语态的大量使用。另一种主张则强调文章应自然，直截了当，要使用第一、第二人称代词。技术杂志 Simulation 在 "作者须知" 中建议在恰当时作者可用第一人称代表他自己，用 you 来称呼读者。下面引用了著名科学家爱因斯坦《相对论》中的一段文字，作者以一种十分自然的谈话口吻娓娓道出。

【例 6】

I am standing in front of a gas range. Standing along-side of each other on the range are two pans so much alike that one may be mistaken for the other. Both are half full of water. I notice that steam is being emitted continuously from one pan, but not from the other. I am surprised at this, even if I have never seen either a gas range or a pan before. But if I now notice something of bluish color under the first pan but not under the other, I cease to be astonished, even if I have never before seen a gas flame. For can only say that this bluish something will cause the emission of the steam, or at least possibly it may do so; If, however, I notice the bluish something in neither case, and if I observe that the one pan continuously emits steam while the other does not, then shall remain astonished and dissatisfied until have discovered some circumstance to which I can attribute the different behavior of the two pans.

从例 6 中可以看出爱因斯坦自然地使用了 I。这种写作手法也越来越多地被人们所采用。

【例 7】

First, you adjust the dial at the top of the machine to the temperature you wish. Then you turn the switch marked "on-off" to start the machine

例 7 是一个操作说明。其中使用了人称代词 you，实际上是告诉人们如何做某事，使用第二人称表达会更加清楚。

（5）正确使用时态

在科技英语写作中，作者要向读者表述研究过程中各项事实、观点产生的时间，因此，正确使用时态就成为十分值得注意的问题。常见的方法有以下几种：

① 表示研究目的时，动词时态一般要使用过去时，这是因为研究的目的是在着手研究时确定的。

【例 8】

The aim of this study was to solve the calculating method of this problem.

② 表示结论时，动词时态一般使用现在时态或现在完成时态，而且以使用现在时为主，除非在强调已获得的成果时方可使用现在完成时。

【例 9】

It is concluded that the principles of the test system allow increased safety and accuracy in hospital drug handling.

③ 表示研究过程中动作或状态时，动词时态一般使用过去时，这是因为研究工作是在撰写论文之前进行的。所以，当在论文中叙述作者做了哪些研究工作、研究工作过程中出现了什么现象、得出什么结果等，所描述的都是过去做过的事，只能使用过去时态。但有时会强调某一行为或状态业已完成或持续到撰写论文时，也会使用现在完成时。

【例 10】

The results of treatment of early gastric carcinoma were analyzed in 65 patients.

【例 11】

We have examined the final results of our experiments and concluded that our expectation have been fulfilled

④ 说明图表的时态一般使用现在时。

【例 12】

Diagrams showing yields are shown in Figure 3. The second column of Table 2 represents the dry weight of tops.

二、结尾句（conclusion sentence）的写作方法

结尾句放在一段的末尾，总结这一段的要点，也称为结论句。结尾句通常与主题句一样包括了该段文字的控制思路。当使用到主题句的控制思路时，在结尾句中所使用的词汇要与主题句中控制部分所使用的词汇有所不同。

【例 13】

Color-blind people have problems that people who perceive color never think about. One very problem is that of traffic signs and signals. Those that are red-green color-blind have trouble seeing top signs found on shady streets because they may not notice them against the leaves. In addition, they have trouble identifying signal lights and must memorize the position of light to know which signal is being given. Perhaps less of a problem is that of dressing. Those who are fashion-conscious avoid brightly colored wardrobe so that they will not wear clashing colors. Perhaps the greatest handicap of color-blind persons is evident when they select occupations. For example, they cannot work as interior decorators, commercial photographers, painters, airline or ship pilots, or railroad engineers. *As most color-blind people cope with their problems, their handicap goes unnoticed.*

例 13 的内容是关于色盲人士所面临的问题。在划线的两个句子中，第 1 句为主题句，第 2 句为结尾句。从两句的内容来看，都涉及一个共同的主题"色盲人士的问题"，所不同之处是在主题句中提出该段文字要展开讨论的问题，扩展方向是"色盲人士与正常人的不同"，而在结尾句中所表述的是在经过对主题句的内容进行充分讨论与扩展后对全段文字进行概括性的总结，"虽然色盲人士与正常人不同，他们会面临很多问题，但他们的这种不同容易被忽视"，其内容不仅与主题句相呼应，还对下一段的内容有所暗示，承上启下。

结尾句要抓住主题句中的关键词，总结一段的要点，可能会用到 to sum up, in short, in conclusion 等词组或词汇。

这里需要指出的是，结尾句并不是在所有的段落文体中均要表示出来。在以举例法（如例 13）、因果法扩展的段落，要有结尾句。有些段落，如以空间顺序法或时间顺序法扩展的段落则可以不要结尾句。

正文的写作技术就此告一段落，然后，我们就要面对科技论文主体写作的最后一个环节——结论的写作。

三、结论部分的写作方法

1. 结论段落的文体特点

结论段中的文字是要将讨论予以结束并对其进行总结。英语科技论文的结尾不外乎起三种作用：①使读者对于文章的内容得到一个全面、清晰、明确的印象；②能激发读者接受文章内容后，进一步探索问题，思考问题；③给读者以强有力的感应，促使读者对讨论的问题或与之相关的问题进行更深入的研究和探讨。

因此，可以说结尾是文章逻辑发展的必然和自然结果。在结尾段中要反映经过分析、综合、推理、判断、比较、实验等研究过程所形成的总的观点。在结尾段中要把文章各部分综合联系起来，突出文章的中心思想。另外，还要注意题目、前言与结尾前后呼应，做到首尾一致。

下面是常见的几个科研论文的结尾的实例。

【例 14】

Conclusion

These two new windmill designs will, of course, have an even more important impact on the energy industry in the future. Although these two designs may never be used on a large scale, their impact may be felt through influencing the design of even more efficient models in the years ahead.

在例 14 中，首先使用了小标题 Conclusion，这是科研论文常用来概括性地表明要研究讨论的内容的一种表达方式。本段除了对文章的主题进行概括性总结外，还在最后一句话中对所研究的工作在未来的可能性进行了必要的预测。另外一个更有说服力的例子就是本章 Part 1 的阅读材料。

【例 15】

Conclusion

Thus, the different constituents of sunlight are treated in different ways as they struggle through the earth's atmosphere. A wave of blue light may be scattered by a dust particle, and turned out of its course, and so on, until finally it enters our eyes by a path as zigzag as that of a flash of lightening. Consequently, the blue waves of the sunlight enter our eyes from all directions. And that is why the sky looks blue.

例 15 则是另一种文体的结尾段，它是回答某一问题的文章结尾。文章的标题是 Why the Sky Looks Blue? 最后一句是对全文所讨论的问题的答案和总结。

总而言之，由于英语科技论文的特点、内容与结构的不同，文章的结尾也千姿百态。

2. 写结论段时应注意的几个问题

写英文的科技论文时必须考虑到英美文化的特点以及英汉文化的差异，在写结尾段时要注意避免下述几个问题。

（1）避免不必要的谦虚

科技论文的特点是实事求是，研究的成果是什么就是什么，既不要吹嘘，也不要过分谦虚，这是英美文化中最忌讳的事。

【例 16】

There must be many mistakes and shortcomings in my paper. I hope you will criticize my paper.

这样的语句一旦出现在结论段中，无论论文的正文所讨论的问题的意义有多大，所取得的研究成果有何等重要，在别人的眼里也是不值得一看的。

（2）在结尾时不要削弱或偏离文章的主题

如在一篇叙述计算机的用途的文章结尾，突然冒出了例 17 这样的句子，就会使人感到与文章的主题相违背，而成了另一个议题。

【例 17】

Of course, computer has its harmful social effects.

显然，应把例 17 作为另一议题的主题句来讨论，而不是对计算机的用途进行总结。这本身也违背了文章内容的一致性与连贯性的要求。

（3）避免使用口语化的表达方式

这是英语科技文章的文体要求所决定的。如果在结论段使用了例 18 这样的表达方式，会使人感到文章成了一个演讲的书面稿。

【例 18】

So I want to close my paper by summarizing the above discussion.

3. 在结论段中常用的表达方式与句型

在结论段中常用的表达方式：

In conclusion…

In summary…

The above account of…

在结尾段中常用的词汇与词组：

thus，hence，therefore，put things together

表示结论的常用句型：

The course program/the experiment described above is…

It is clear from the foregoing discussion that …

Demonstration/ Illustration/ Classification/ Comparison/ Contrast gives/shows…

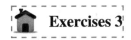 Exercises 3

Have you written something about your daily discharge of MSW? Please use developing sentences to enrich your topic sentence in every paragraphs. Finally, draw a conclusion of the article.

 Expanding Reading

Ecological Risk Assessment

An ecological risk assessmentis the process for evaluating how likely it is that the environment might be impacted as a result of exposure to one or more environmental stressors, such as chemicals, land-use change, disease, and invasive species.

An ecological risk assessment includes three phases, but begins with Planning:
- identify risk management goals and options;
- identify the natural resources of concern;
- reach agreement on scope and complexity of the assessment;and
- decide on team member roles.

Phase 1　Problem Formulation

The risk assessor(s) gathers information to determine which plants and animals are or might be at risk and in need of protection. Based on the Planning results, they specify:
- the scope of the assessment in time and space;
- the environmental stressors of concern;
- the endpoints to be evaluated (e.g., continued existence of a fishery population, fish species diversity in lakes, sustainable forest habitat); and
- which measures, models, and type of data will be used to assess risks to those endpoints.

Problem formulation concludes with an Analysis plan.

Phase 2 Analysis

Two components of the analysis phase are exposure and effects assessments. In the exposure assessment, the risk assessor determines which plants and animals are or are likely to be exposed to each environmental stressor and to what degree. In the effects assessment, the risk assessor reviews available research on the relationship between exposure level and possible adverse effects on plants and animals. They may also review evidence of existing harmful ecological effects.

Phase 3 Risk Characterization

Risk characterization includes two major components—risk estimation and risk description. "Risk estimation" includes the estimated or measured exposure level for each stressor and plant or animal population, community, or ecosystem of concern; and the data on expected effects for that group for the exposure level. "Risk description" provides information important for interpreting the risk results. This includes:

- whether harmful effects are expected on the plants and animals of concern;
- relevant qualitative comparisons; and
- how uncertainties (data gaps and natural variation) might affect the assessment.

(*https://www.epa.gov/risk/ecological-risk-assessment*)

Unit 5 Environmental Management and Environmental Impact Assessment

Lesson 2 Environmental Impact Assessment (EIA)

Part 1 Reading

1 Purpose

EIA is intended to identify the impacts (both beneficial and adverse) of a proposed public or private development activity. Often, the focus is dominantly environmental (biophysical); but practice also addresses social and economic aspects. EIA is mainly used at the level of specific developments and projects such as dams, industrial plants, transport infrastructure (e.g. airport runways and roads), farm enterprises, and natural resource exploitation (e.g. sand extraction). Strategic Environmental Assessment (SEA)[①] is a sister tool applied upstream at the level of policies, plans and programmes. Like SEA, EIA is most valuable when applied early in the planning process for a project as a support to decision-making. It provides a means to identify the most environmentally suitable option at an early stage, the best practicable environmental option and alternatives to the proposed initiative; and thus avoid or minimize potentially damaging and costly negative impacts, and maximize positive impacts.

2 Background Facts

EIA was first introduced in the USA under the National Environmental Policy Act (1969)[②]. Since then it has evolved and a variety of offshoot assessment techniques have emerged (focusing, for example on social, biodiversity, environmental health and cumulative effects and risk) acting as a broader impact assessment toolkit. Most countries have now introduced formal EIA systems, usually under dedicated environmental legislation, and have introduced EIA regulations (and often regulatory bodies) specifying when and for which developments an EIA is required, institutional responsibilities and procedures, and specific steps and processes to be followed.

3 Brief Description of the Main Steps Involved in Application of the Tool

Key stages in the Environmental Assessment process include: screening, alternatives, preliminary assessment, scoping, mitigation, main EIA study and environmental impact statement, review and monitoring. These need to be managed so that they provide information to decision-makers at every stage of the project planning cycle.

4 Steps in EIA

(1) Screening (usually by an EIA Authority, or using published checklists) —to decide whether an EIA is required and focus resources on projects most likely to have significant impact, those where impacts are uncertain and those where environmental management input is likely to be required. Official EIA guidelines usually contain lists or schedules specifying which developments require an EIA (e.g. always, or in particular circumstance).

(2) Consideration of possible alternatives (demand, activity, location, process & design, scheduling, inputs, "no project") should be undertaken before a choice is made. Some projects be site specific (e.g. in mining, extraction can only occur were a mineral is sited). In such cases the EIA might focus more on measures such as scale, mitigating measures and traffic management. Projects promoted by public sector agencies are more likely to consider alternative sites or routes for development than private sector initiatives where the early need to acquire options or purchase land strongly influences development location.

(3) Preliminary assessment—where screening suggests further assessment is needed or if there is uncertainty about the nature of potential impacts. Uses rapid assessment techniques, but provides sufficient detail to identify key impacts, their magnitude and significance, and evaluate their importance for decision-making. Indicates if a full EIA is needed—involving the following steps.

(4) Scoping — a "narrowing" process usually undertaken by an assessment team to identify key issues of concern at an early stage in the planning process and guide the development of terms of reference for the EIA. It aids site selection, identifies possible alternatives, and avoids delays due to having to assess previously unidentified possible impacts. Scoping should involve all interested parties such as the proponent and planning or environmental agencies and members of the public. The results determine the scope, depth and terms of reference to be dressed within an Environmental Impact Statement (see below). Once the site for development has been selected, the number of issues usually decreases and attention to detail increases.

(5) Main EIA study—building on and deepening the preceding steps to predict the extent an magnitude of impacts and determine their significance. A variety of methods can be used including: checklists, questionnaires, matrices, overlays, networks, models and simulations. The study should incorporate consideration of mitigating measures—reviewing the action proposed/taken to prevent, avoid or minimize actual or potential significant adverse effects of a project, e.g. abandoning or

modifying a proposal, or substituting techniques using BATNEEC (Best Available Technology Not Entailing Excessive Costs)③ such as pollution abatement techniques to reduce emissions to legal limits.

If the uncertainties are great, with the possibility of grave consequences and no mitigating measures then the proposed development should be rejected. If there are uncertainties that might be reduced by further studies, then an application can be deferred pending until further studies. Where mitigation is inappropriate, compensation may be an option.

An Environmental Impact Statement (EIS)④ is a comprehensive document that reports the findings of the EIA and now often required by law before a new project can proceed. A typical EIS, usually prepared by the project on behalf of the proponent (usually by consultants), focuses on the issues most relevant to decision-making. It can be broken down into three parts with different levels of detail:

- Volume 1—a comprehensive and concise document drawing together all relevant information regarding the development project.
- Non-Technical Summary (NTS)—a brief report of volume 1 in non-technical language that can easily be understood by the public.
- Volume 2—a volume that contains a detailed assessment of the significant environmental effects (Not necessary when there are no significant effects either before or after mitigation).

Alternative communication approaches by also be appropriate where literacy or social/cultural barriers prevent local people accessing the EIS (e.g. local language videos, presentations, radio programmes, meetings and workshops).

(6) Review—to assess the adequacy of the EIA to decision-making and consider its implications for project implementation (in some countries, such review is a formal and independent process).

(7) Monitoring of project implementation and operation (including decommissioning), and eventually an audit of the project after its completion.

5 Expected Outputs

- An Environmental Impact Statement that provides clear, understandable, relevant information to influence the final decision on the development project.
- A better development project (minimized negative impacts, maximized positive impacts, optimal location, best alternative selected, etc.)

6 Basic Requirements

For a major project, an EIA may take considerable time, manpower and resources. The first four stages are very important to determine the required extent and focus of the EIA.

7   Data

Prediction of impacts relies on data from a variety of sources: physical, biological and sociological. Its quality will often impose constraints on accuracy and reliability of predictions. Where data is limited, qualitative techniques will need to be used.

Skills and Capacity—often a multidisciplinary team is required-particularly where scoping indicates the existence of multiple or complex issues.

Pros (main advantages) and Cons (main constraints in use and results)

- EIA often focuses on biophysical issues (often a fault of poor terms of reference).
- Where environment, social and economic aspects are addressed, they are not always addressed in an integrated way.
- EIA provides an opportunity to learn from experience of similar projects and avoids the (often high) cost of subsequently mitigating unforeseen negative and damaging impacts.
- EIA improves long-term viability of many projects.

8   Case Study: EIA Mkuze River Crossing to Phelendaba, South Africa

A relatively small EIA of a proposed road development in Maputaland, South Africa was conducted in 1999 for the South African Roads Agency, Department of Transport. The road formed a key infrastructural component of a Spatial Development Initiative aiming to provide a direct link between northern KwaZulu-Natal and Mozambique to encourage rapid investment and convert the area into an internationally competitive zone of economic activity and growth. The project aimed to tar the road, upgrade river crossings and construct a new crossing over a swamp. The undeveloped area dominated by subsistence agriculture had high levels of poverty and unemployment but high biodiversity value and high eco-tourism potential.

An extended scoping study proved adequate for decision-making, despite the complexity of issues. It proved unnecessary to undertake an intensive, detailed EIA that would have had significant time and resource implications.

The study involved extensive stake holder participation. Because of the eco-tourism potential of pans in the area, the KwaZulu-Natal Nature Conservation Service (KZNNCS) proposed an alternative routing for the roadway (the western alignment), arguing that this would provide greater access to the Kwa-Jobe Tribal Authority-an extremely poor community.

The existing route was shorter and therefore cheaper, and most of its alignment was already cleared, so linear developments and other disturbances already existed. But it passed through a state forest with hazards to game and game hazards to traffic, opportunistic poaching and noise from the road.

The alternative western alignment required clearing 140 ha of mature Sand Forest and could open access to uncontrolled woodcutting. But benefits included expanding the width of a migration corridor between the Mkuze Game Reserve and the Sodwana.

Words and Expressions

infrastructure	[ˈɪnfrəstrʌktʃər]	n. （国家或机构的）基础设施，基础建设
extraction	[ɪkˈstrækʃn]	n. 提取，提炼，拔出，开采
initiative	[ɪˈnɪʃətɪv]	n. 倡议，新方案，主动性，积极性，自发性，掌握
offshoot	[ˈɔːfʃuːt]	n. 分支，（尤指）分支机构
toolkit	[ˈtuːlkɪt]	n. 工具箱，工具包，配套软件
legislation	[ˌledʒɪsˈleɪʃn]	n. 法规，法律，立法，制定法律
regulatory	[ˈreɡjələtɔːrɪ]	adj. 具有监管权的，监管的
screening	[ˈskriːnɪŋ]	n. 筛选（环境影响评价因子）
preliminary assessment		n. 初步评估
scoping	[ˈskoʊpɪŋ]	n. 评价范围界定
mitigation	[mɪtɪˈɡeɪʃn]	n. 减缓措施
no project		n. 零方案
substitute	[ˈsʌbstɪtuːt]	v.（以……）代替，取代
abatement	[əˈbeɪtmənt]	n. 减弱，减轻，减少
defer	[dɪˈfɜː]	v. 推迟，延缓，展期
pending	[ˈpendɪŋ]	adj. 待定，待决，即将发生的；v. 吊着，悬而未决，待决
proponent	[prəˈpoʊnənt]	n. 倡导者，支持者，拥护者
decommissioning	[ˌdiːkəˈmɪʃnɪŋ]	v. 正式停止使用（武器、核电站等）
multidisciplinary	[ˌmʌltiˈdɪsəplənerɪ]	adj. （涉及）多门学科的
alignment	[əˈlaɪnmənt]	n. （国家、团体间的）结盟
poach	[poʊtʃ]	v. 偷猎，偷捕

Notes

① Strategic Environmental Assessment (SEA)战略环境评价

② the National Environmental Policy Act (1969)美国《国家环境政策法》（1969），是一部在世界环境法立法史上占有重要地位的法律，它以一系列立法创新为美国当代环境法制建设奠定了基础，也是世界上第一个引入环境影响评价制度的法律。

③ Best Available Technology Not Entailing Excessive Costs (BATNEEC)最佳可用技术，无

需额外成本。

④ Environmental Impact Statement (EIS) 环境影响报告书

Exercises 1

1. According to the reading material, chose the best answer(s) from the options.

(1) EIA is intended to identify the impacts (both beneficial and adverse) of a proposed public or private development activity. Often, the focus is dominantly _____ , _____ and _____ aspects.

  A. environmental B. biophysical C. social D. economic

(2) EIA was first introduced in the USA under the National Environmental Policy Act in _____.

  A. 1996 B. 1969 C. 1999 D. 1699

(3) Key stages in the Environmental Assessment process include _____.

  A. screening, alternatives, preliminary assessment B. scoping, mitigation

  C. main EIA study D. environmental impact statement, review and monitoring

(4) Pros and Cons of EIA include _____.

  A. EIA often focuses on biophysical issues (often a fault of poor terms of reference)

  B. Where environment, social and economic aspects are addressed, they are not always addressed in an integrated way

  C. EIA provides an opportunity to learn from experience of similar projects and avoids the (often high) cost of subsequently mitigating unforeseen negative and damaging impacts

  D. EIA improves long-term viability of many projects

2. List steps involved in EIA application in English.

3. List the methods which can be used to predicate extent and magnitude of environmental impact for the proposed development activities in English.

Part 2 Translation

英语翻译中从句的翻译（二）——状语从句

状语从句在整个句子中用来修饰主句中的动词、形容词或副词。英语中的状语从句形式多样、变化繁多，是一个比较复杂的语言现象。从类别上来说，英汉状语从句可以分为九大类，即时间状语从句、地点状语从句、原因状语从句、结果状语从句、条件状语从句、方式状语从句、目的状语从句、让步状语从句、比较状语从句等。从语序上来说，英语中状语从句的位置比较灵活，可放在主句的前面、中间或后面；而汉语中状语成分的位置较为固定，一般放在主句的前面。

英语和汉语分别隶属不同语系，前者属于印欧语系，具有形合式的特点，而后者则属于汉藏语系，具有意合式的特点。换言之，汉语语句在表达上主要靠词序或语序，少用连词，而英语则不同，它必须要有相应的连词加以连接，但在翻译的过程中，往往不必译出连词。例如：

(1) Because heat does not take up any room and it does not weigh anything, it is not a material.

译文1：热不占有任何空间，也不具有什么重量，因此，它不是物质。

译文2：因为热不占有任何空间，也不具有什么重量，它不是物质。

(2) As the bit drills down, pipe is added length after length.

译文1：随着钻头往下钻，管子一段一段地加接上去。

译文2：当钻头往下钻时，管子一段一段地加接上去。

比较以上两个例子，译文1和译文2孰优孰劣，不言自明。

一、时间状语从句

英语中的时间状语从句，一般置于主句之前或之后，而在汉语中，通常置于主句之前。翻译时，往往采用以下几种处理方法，即顺序译法、逆序译法、分译译法、转换译法。例如：

(1) Whenever we think hard, our brain produces special transport chemicals.

译文：当我们思考时，我们的大脑产生特殊的传输化学物质。（顺序译法）

(2) Hardly had the operator pressed the button when all the electric machines began to work.

译文：操作人员一按电钮，所有电机都开始运作。（逆序译法）

(3) The earth turns round its axis as it travels about the sun.

译文：地球一边绕太阳转，一边绕地轴自转。（分译译法）

(4) Turn off the switch when anything goes wrong with the machine.

译文：如果机器发生故障，（那么）请把电闸关上。（转换译法）

二、地点状语从句

地点状语从句一般是由 where，wherever 等连词引导，位置相对比较灵活，可放在主句之前也可放在主句之后，有时，地点状语从句表达的意思是条件句或结果状语从句的意思。翻译时，也可灵活处理，一般可采用顺序译法、逆序译法和转换译法等翻译方法，例如：

(1) We use insulators to prevent electrical charges from going where they are not wanted.

译文：我们使用绝缘体是为了防止电荷跑到不需要的地方去。（顺序译法）

(2) A signal will be shown wherever anything wrong occurs in the control system.

译文：无论控制系统什么部位出故障，都会给出信号。（逆序译法）

(3) Variations and permutations of these procedures are employed, particularly where equipment on hand has been modified for mask manufacture.

译文：这些程序可以加以改变或重新排列，特别是现有的设备为制造膜而已经被修改了。（地点状语从句转译为目的状语从句）

三、原因状语从句

英语中原因状语从句常常是由 because, since, for 和 as 等连词引导，往往位于主句之后，有时也位于主句之前。在汉语表达习惯中，一般先说明原因，再说结果，这样才符合逻辑。所以，在翻译时，往往被译成"由于……""因为……"等句式。在有些情况下，受到印欧语言的影响，某些原因状语从句放在后面，只是起到补充说明作用。以下来具体分析几种翻译方法，例如：

(1) The material first used was copper for the reason that it is easily obtained in its pure state.

译文：最先使用的材料是铜，因为易于制取纯铜。（顺序译法）

(2) Some sulphur dioxide is liberated when coal, heavy oil and gas burn, because they all contain sulphur compounds.

译文：因为煤、重油和煤气都含有硫化物，所以它们燃烧时会释放一些二氧化硫。（逆序译法）

(3) Because he was convinced of the accuracy of this fact, he stuck to his opinion.

译文：他深信这一事实正确可靠，故而坚持己见。（转换译法）

四、结果状语从句

英汉两种语言中结果状语从句都放在主句之后，由连词 that 或 so...that, such...that 等结构引导。多数情况下，采取顺序译法处理就可以了，但也不要过于拘泥，通通将其译为"因而""因此""如此……以致""致使……"等句式。一句话，欲得佳译，还需译者灵活处理。例如：

(1) The sun is so much bigger than the earth that it would take over a million earths to fill a ball as big as the sun.

译文：太阳比地球大多了，所以需要一百多万个地球才能填满像太阳那么大的球体。（顺序译法）

(2) The computer should be designed in such a fashion that it can find the right bit of knowledge at the right times.

译文：计算机的设计应使它能在适当的时间找到适当的知识。（顺序译法）

(3) Some stars are so far away that their light rays must travel for thousands of years to reach us.

译文：有些恒星十分遥远，它们的光线要经过几千年才能到达地球。（转换译法）

(4) Electricity is such an important energy that modern industry couldn't develop without it.

译文：电是一种非常重要的能源，没有它，现代工业就不能发展。（转换译法）

五、条件状语从句

英语中的条件状语从句主要由 if, unless, as/ so long as, on condition that 等连接词引导，其在逻辑关系上与汉语的假设复句或条件复句大体一致。一般而言，英语中的条件状语从句位于主句之前或之后，但绝大多数位于主句之前，而在汉语中，它往往位于复句的前部。在某些情况下，为了强调说明或附加补充，也可以将其置于主句之后，翻译时，也可依此译出。例如：

(1) If something has the ability to adjust itself to the environment, we say it has intelligence.

译文：如果某物具有适应环境的能力，我们就说它具有智慧。（顺序译法）

(2) If too much current flowed through an ammeter it would cause the voltage to fall when it was connected, and the measured voltage would be wrong.

译文：如果过多的电流流过电流表，在连接时会导致电压下降，测量的电压就会出错。（顺序译法）

(3) Electricity would be of very little service if we were obliged to depend on the momentary flow.

译文：在我们需要依靠瞬时电流时，电力就没有多大用处。（条件状语从句转译为时间状语从句）

此外，还有一种特殊的条件句，就是英语中的非真实条件句，也称作虚拟语气。这类句子位于句首，省略 if 引导词，却把 were, had, should 和 could 等词提到句首主语之前，形成一种倒装语序。对于这类句子，在翻译时，也可参照由连词 if 引导的条件句的处理方法，常常采用顺序译法。例如：

(4) Were there no electric pressure in a semiconductor the electron flow would not take place in it.

译文：如果半导体中没有电压，其内部就不会产生电流。

(5) Had the French inventor tried to model his plane on every detail of a bird's wing, he would not have succeeded in flying it.

译文：要是那个法国发明家模仿鸟翼的每个细节来试造他的飞机的话，他就不能使它飞起来。

这类条件句在科技英文中俯拾皆是，在此就不一一列举。

六、方式状语从句

方式状语从句是英语中较为特别的一种状语从句，它可以位于主句之前，也可以位于主句之后，一般由 as,（just）as...so..., as if, as though 等引导。as,（just）as..., so...等引导词引导的方式状语从句通常位于主句后，但在（just）as...so...结构中会位于句首，这时 as 从句带有比喻的含义，意思为"正如……""就像……"，多用于较为正式的文体当中。as if、as though 两者的意义和用法相同，引出的状语从句谓语多用虚拟语气，表示与事实相反，有时也用陈述语气，表示所说情况是事实或实现的可能性较大，常常译为"仿佛……似的""好像……似的"。翻译时，需要对这些引导词加以区分，正确译出其意义。例如：

(1) Each of two wires carrying currents has a magnetic field, as if it were a magnet.

译文：通电的两根导线每根都有一个磁场，就好像是一块磁铁一样。（顺序译法）

(2) Just as all living things need air, water and sunlight, so plants need them.

译文：植物像所有生物一样也需要空气、水和阳光。（逆序译法）

(3) The computer as it seems to play the role of a human brain is often called an electronic brain.

译文：由于计算机的作用类似于人脑，所以常被称为电脑。（方式状语从句转译为原因状语从句）

七、目的状语从句

英语中的目的状语从句与汉语中的表达习惯基本相同，一般来说它位于主句的后面，常常由 in order that、so that 等结构引导，翻译时，往往处理为"以便……""以免……""为的是……""为了要……"等句式。此外，在由 so that 引导的表示要达到某种目的的一种愿望的从句中，其谓语里往往含有诸如 may、might、can/could/should 等的情态动词，否则就属于表示一种结果的结果状语从句，试比较以下几个例句。例如：

(1) Many machine parts are now made of plastics instead of metals (so) that their weight may be decreased.

译文：现在许多机器部件不用金属而用塑料制造，为的是减轻重量。（目的状语从句）

(2) The negative work should be maintained at as low a value as possible for a given pressure ratio, so that a maximum amount of useful work may be obtained.

译文：在给定的压力比下，负功应当保持在尽可能低的值，以便可以获得最大的有用功。（目的状语从句）

(3) Reflectors are located so closely to each other that echoes overlap, resulting in the single continuous echo signal train.

译文：发射器彼此密切相连，因此反射波重叠，形成单一连续的反射信号串。（结果状语从句）

以下来具体分析目的状语从句的翻译方法，例如：

(4) Data processing involves filtering and summarizing data so that underlying patterns can be perceived.

译文：数据处理包括筛选和汇总数据，以便可以探知数据的基础结构。（顺序译法）

(5) He pushed open the door gently and stole out of the room for fear that he should awake her.

译文：为了不惊醒她，他轻轻推开房门，悄悄地溜了出去。（逆序译法）

八、让步状语从句

英语中的让步状语从句往往由 while，although，though，even though 等连词引导，也可以由"no matter+疑问词"或"疑问词+后缀 ever"或"形容词+as+倒装句子"构成。一般情况下，由 though 和 although 引导的让步状语从句，后面的从句中不能出现 but 一词，但是 though 和 yet 可连用；由 as 和 though 引导的让步从句必须将表语或状语（如形容词、副词、分词等）提前。翻译时，常常译为"虽然/尽管/即使……但……""不管/无论……"等意义的句式，常采用顺序译法、逆序译法和变序译法。例如：

(1) Although he seems hearty and outgoing in public, Mr. Cooks is a withdrawn, introverted man.

译文：虽然库克斯先生在公共场合热情而开朗，但他却是一个孤僻的、性格内向的人。（顺序译法）

(2) All electronic computers consist of five units although they are of different kinds.

译文：尽管种类各异，但所有电子计算机都由五大部分组成。（逆序译法）

(3) The normal temperature for a human being, no matter in what part of the world he lives, is about 37 ℃.

译文：不管生活在世界上什么地方，人的正常体温都在 37 ℃左右。（变序译法）

九、比较状语从句

英语中的比较状语从句通常是由 as... as...，...than，the more... the more...，not so...as... 引导，有些句子成分通常可省略。这类从句和大多数状语从句不同，它们不修饰动词，而是修饰 as/so/less/more 等副词，或其他比较级的词，如 taller，harder 等。这里讨论的是它们作为状语从句的情况。一般而言，英汉语中的比较状语从句位于主句之后，往往译为"比……""和……一样"，可采用顺序译法和变序译法。例如：

(1) The faster the gas rushes out, the faster the rocket moves. （顺序译法）

译文：气体喷出速度越快，火箭的运行速度也就越快。

(2) The better conductor a substance is, the less is its resistance. （顺序译法）

译文：物质的导电性能越好，其电阻越小。

(3) The UK will have more flexibility than before over trading-off time, whole-life cost and performance.

译文：英国在交易时间、全寿命费用和性能等方面将具有前所未有的灵活性。（变序译法）

Exercises 2

Translate the following paragraph into Chinese.

(1) Projects promoted by public sector agencies are more likely to consider alternative sites or

routes for development than private sector initiatives where the early need to acquire options or purchase land strongly influences development location.

(2) In such cases the EIA might focus more on measures such as scale, mitigating measures and traffic management. Projects promoted by public sector agencies are more likely to consider alternative sites or routes for development than private sector initiatives where the early need to acquire options or purchase land strongly influences development location.

(3) Key stages in the Environmental Assessment process include: screening, alternatives, preliminary assessment, scoping, mitigation, main ELA study and environmental impact statement, review and monitoring. These need to be managed so that they provide information to decision-makers at every stage of the project planning cycle.

(4) Since then it has evolved and a variety of offshoot assessment techniques have emerged (focusing, for example on social, biodiversity, environmental health and cumulative effects and risk) acting as a broader impact assessment toolkit.

(5) Projects promoted by public sector agencies are more likely to consider alternative sites or routes for development than private sector initiatives where the early need to acquire options or purchase land strongly influences development location.

Part 3　Writing

学术论文的撰写（九）——摘要和关键词的写作技巧

一、摘　要

摘要（abstract）也称为内容提要，美国国家标准（ASNI）定义摘要是"一篇精确代表文献内容的简短文字"。摘要的英文术语原来有两个词汇，一个是"abstract"，一个是"summary"，现在越来越多的用法是"abstract"，尤其是放在索引资料中一律要用"abstract"这个术语。"summary"现在更多的是用于对文章主要内容的陈述，放在文章的结尾。通常在科技论文中必须附有摘要，其位置应放在正文之前。

1. 摘要的功能与种类

摘要的用途很广，除了论文中要有摘要外，平时在阅读报道、书籍、报告时都可作摘要，以便记住要点，保存待用。在科研工作中写报告时要有摘要，便于专家和有关人员阅读，了解和评估该项科研工作的情况与结果。

在英语科技论文中，摘要是为了便于学术交流。论文摘要常被专业期刊文献杂志编入索引资料或文献刊物，如美国的 SCI 和 EI 等就属于这类刊物。这两种文摘资料性期刊为广大科研工作者带来了许多便利。科技人员阅读科技文章时，绝大多数首先阅读的是论文的摘要，然后再根据摘要的提示确定是否要继续往下读。此外，参加学术会议，一般要首先提供论文摘要，以期通过会议论文审议组的审定，取得参加会议的资格以及宣读论文的资格。在参加国际学术会议期间，与会者也可根据会议所提供的论文目录中的摘要来决定参加哪一个专业小组的讨论。所以，科技工作者要学会写英文摘要。

摘要分为两类，一类是说明性摘要（descriptive abstract），一类是资料性摘要（informative abstract）。如果没有特殊要求，科研人员一般使用的是资料性摘要。现分别对两种摘要进行叙述。

（1）说明性摘要

说明性摘要主要是说明文章的内容，提供文章的信息，概述论文的目的、方法和所涉及的范围，对论文的论题进行说明。所以，当读者读到这类摘要时，便可决定是否有必要阅读整篇论文。因此，说明性摘要的篇幅较短，有时短得只有一句话。下面是一个说明性摘要的例子，请注意其语言特点。

【例1】

This paper presents an analysis of principles of magnetic refrigeration with application to air conditioning. A comparison with conventional evaporation condensation gas cycle device is presented. Conclusions concerning the applicability of magnetic refrigeration to air conditioning are made.

这个说明性摘要只用了三句话就完成了写作。它的内容实际上就是原文的大框架，没有涉及具体的方法与步骤，不是对具体的步骤的简要总结与对实际内容的概括与浓缩，提供给读者的只是一个题目、一个过程和结果，而在过程和结果中没有实际的内容。

在说明性摘要中常用于表示论文所涉及的范围的词及词组有 discuss, examine, review, outline, present, put forward, make, reveal, deal with 等。

（2）资料性摘要

资料性摘要不同于说明性摘要，它是原文内容要点的具体总结，其内容包括目的、方法、论文所涉及的范围、提供结果或者发现；提供结论或对下一步的工作提出建议。下面是一个资料性摘要的实例，请与例1比较，注意其内容、结构上的差别。

【例2】

The science taught in the classroom should be reasonably up-to-date. What is taught should place emphasis first on principles and major concepts of science rather than on the applications of scientific knowledge. The instructional techniques comprise laboratory work which is introduced in such a manner as to emphasize science as a process—to reveal through practice that science involves inquiry, discovery, and experimentation.

It is suggested that college science programs should be revised with a view to preparing teachers to handle science in secondary and elementary schools.

例2是资料性摘要，它对论文的要点进行了概括，提出了实验方法并对今后的研究提出了建议，这个摘要有实际的内容。读者在阅读这种摘要时，就可以对论文的内容有个大概的、较为清楚的了解。所以在资料性摘要中所涉及的是论文的主题研究中所采用的方法、主要研究结果及数据、研究的结论和对开展下一步研究工作的建议等内容。通常，这一类摘要反映了论文的基本面貌，能够代替阅读论文全文。

在资料性摘要中常用于表示论文所涉及的范围的词及词组：in terms of，in view of, considering, referring to, as far as…, (be) concerned with 等。

在写资料性摘要时常用的表示引用文章的主题的词组与句型：

According to the article/report/thesis/paper;

The author maintains that;

This paper describes/shows/provides/discusses/reports/investigates/deals with;

… is concerned with…;

the aim of this investigation (be)…;

The paper is aimed at…;

The paper is limited to…;

… (be)analyzed and evaluated 等。

表示研究方法的词组与句型：

The experiment (be) performed by using…;

The approach (be) based on…;

…(be)based on the use of…;

(be)determined by…;

(be)identified by…;

(be)observed by…;

(be)studied by measuring…;

(be)characterized by…; analysis of…by;

(be)revealed by…等。

表示研究结果和结论的常用句型：

It is concluded that…;

The results are presented in the form of…;

It is suggested that…;

The results confirm that…;

The results (be) summarized as follows…;

The comparison concludes that…;

These findings regarding/suggest that…;

…as has been noted;

To sum up;

In the summary;

In brief/short/a word 等。

2. 摘要的特点

根据联合国教科文组织的规定:"全世界公开发表的科技论文,不管用何种文字写成,都必须附有一篇简练的英文摘要。"那么,对摘要的内容究竟要简练到何种程度才算合乎要求呢?

学术论文摘要的长短一般为正文字数的 2%~3%。国际标准化组织建议不少于 250 个词,最多不超过 500 个词。美国化学文献、医学文摘规定在 200 个词以内。

一篇好的论文摘要需简明扼要,它应包括研究目的、研究对象、研究方法、研究结果、所得结论、结论的适应范围等六项内容。其中,研究的对象与结果是每篇摘要必不可缺的内容,其他项目可按论文的具体内容灵活运用。举例来说,《化学文献》(*Chemical abstracts*)是专供化学工作者使用的一份文摘期刊。从该文献中可以发现各国最有影响的化学杂志对摘要写作有一些共同的要求,归纳起来有三条:

(1)要说明实验或论证中新观察到的事实、结论。若有可能,还要说明新的理论、处理方法及过程、仪器、技术等要点。

(2)要说明新化合物的名称、新数据,包括物理常数等。如不能对其说明,也要提到这些内容。提到新的项目和新的观察是很重要的。

(3)在说明实验结果时,要说明所采用的方法,如果是新方法,还要简述其基本原理、操作范围和准确程度。

下面是一份关于双流体模型的论文摘要。

【例3】

The two-fluid model of turbulence proposed in 1982 describes various states of fluids at the same position of space. The present paper gives a brief introduction to the basic principle of this model and lays emphasis on the new idea about the interaction between the two fluids and corresponding mathematic models, including the transformation of the turbulent fluid into the non-turbulent fluid by decay and dissipation of turbulence. In a turbulent free jet, the existence of heat transfer between combustion products and unburnt combustible mixture in a zone of premixed combustion, and the intensification of chemical reaction by the rising temperature of combustible mixture. Predictions of the intermittency of turbulence in a turbulent free jet and of a turbulent premixed combustion are likely to be improved with this model.

例 3 中第一句话对双流体模型进行了概述,然后简述了新的构想和相应的数学模型及基本内容,以及经过改进的双流体模型在运用中所取得的结果。

3. 摘要的主要内容及文体要求

(1)摘要的主要内容

科技论文摘要的主要内容在上文中已经提及,资料性论文摘要应包括研究目的(对论文题目的说明)、研究对象(所研究的课题)、研究方法(实验手段、实验设备、实验条件和实验过程)、研究结果(新的发现或所取得的研究成果)、所得结论(根据新发现和所取得的成果得出的研究结论以及对下一步科研工作提出的建议）、结论的适用范围（对所取得的成果的科学性与适用性进行界定）和关键词（能体现论文内容与主题）等七项内容。其中，研究的对象、结果和关键词是每篇摘要必不可缺的内容，其他项目可按论文的具体内容灵活运用。下面就是一个内容比较全面的论文摘要的实例。

【例4】

The article takes up some of the issues identified by Douglas (2000) as problematic for Language for Specific Purposes (LSP) testing, making reference to a number of performance-based instruments designed to assess the language proficiency of teachers or intending teachers. The instruments referred to include proficiency tests for teachers of Italian as a foreign language in Australia (Elder, 1994) and for trainee teachers using a foreign language (in this case English) as medium for teaching school subjects such as mathematics and science in Australian secondary schools (Elder, 1993b; Viete, 1998).

The first problem addressed in the article has to do with specificity; how does one define the domain of teacher proficiency and is it distinguishable from other areas of professional competence or, indeed, from what is often referred to as "general" language proficiency? The second problem has to do with the vexed issue of authenticity: what constitutes appropriate task design on a teacher-specific instrument and to what extent can "teacher-like" language be elicited from candidates in the very artificial environment of a test? The third issue pertains to the role of nonlanguage factors (such as strategic competence or teaching skills) which may affect a candidate's response to any appropriately contextutalized test-task and whether these factors can or should be assessed independently of the purely linguistic qualities of the test performance.

All of these problems are about blurred boundaries, between and within real world domains of language use, between the test and the nontest situation, and between the components of ability to knowledge measured by the test. It is argued that these blurred boundaries are an indication of the indeterminacy of LSP, as currently conceptualized, as an approach to test development.

例4由3段文字组成，共计280个词汇。这个摘要，包括了研究的目的、要解决的问题、所采用的方法、论文所涉及的范围以及研究的结论。

（2）摘要的文体要求

摘要应遵循准确、简明、清楚的原则。准确是指内容上要忠实于原文，内容是标题的扩充，是全文的高度概括，所以在摘要中不能加入自己的评价。简明是摘要文体要求的一个重要内容，在摘要中要使用正规英语、标准术语，避免使用缩写词汇，也不能使用图和表格。清楚是指使用简洁的正规英语将文章的论题、论点、实验方法、实验结果用有限的字数表达出来，要做到不遗漏不重复、不使用带有感情色彩和意义不确定的词汇，也不使用祈使句和感叹句。

①在摘要中一般要使用第三人称单数做主语，以体现其客观性，如 This paper described…，The article takes up…，…is discussed in this paper，…is studied 等。

②如何在摘要中使用正确的时态是尤其要注意的。一般来说，摘要中的时态应与论文相

应部分中的时态保持一致。其基本的要求如下：涉及背景信息的句子通常使用一般现在时；主要的行为一般应使用过去时或现在完成时；对方法的叙述应该使用过去时；对结果的表述应使用过去时；结论则要使用一般现在时。

【例 5】

This paper describes a reflective noticing activity in which pairs of adult learners of English for Academic Purposes transcribed their own performances of a routine classroom speaking task. Working collaboratively, they then discussed and edited the transcripts, making a large number of changes, which were overwhelmingly for the better. These edited transcripts were passed on to the teacher, who made further corrections and reformulations, and then discussed the changes with the learners. Analysis of the process and product of these cycles of work suggests that collaborative transcribing and editing can encourage learners to focus on form in their output in a relatively natural way. It also underlines the role of the teacher in this sort of post-task intervention, especially in the area of vocabulary.

在例 5 中，对主要的活动，如 transcribed their own performances，discuss and edited the transcripts, these edited transcripts were passed on to, made further corrections，discussed the changes 都是用动词的过去时表示的。

在表示结论"Analysis of the process and product of these cycles of work suggests that collaborative transcribing and editing can encourage learners to focus on form in their output in a relatively natural way. It also underlines the role of the teacher in this sort of post-task intervention especially in the area of vocabulary."时，用的是一般现在时。

4. 摘要的写作技巧与注意事项

根据摘要的写作要求与文体特点，它应是论文的高度浓缩。摘要写作应掌握以下几种方法：

（1）精练。这是摘要的文体特征所要求的。如果当前的研究是在他人研究工作基础上发展出来的，则必须说明研究的根据，也需要提到别人的研究。遇到这种情况，则只需要用 On the basis of…'s research/conclusion/theory/methods of…，或者用 According to…'s theory 等表达方式即可。不要使用繁杂的表达方式，如 After an extended series of trials over a period of several month following…theory/ conclusion 等。

（2）抓住重点，完整表达。摘要应充分表达论文的主要内容，所以在撰写论文摘要时，要明确哪些内容是论文的重点。下面介绍一种写摘要时确定重点内容的方法。

【例 6】

原文	重点注记
When and how to cut your losses? Half the skill in getting ahead on the career front <u>is knowing when to move on</u>. In everyone's life there comes a moment when they should make the break—the world is	important to know when to change jobs

full of has-beens-who, perhaps, <u>didn't have the courage to take a chance</u> when that chance came. It pays to constant reassess where you stand. A good stock question to ask yourself is "<u>Where am I going to be this time next year</u>, if I stay in the same job? " Each career has a different time of scale. The sales scene moves fast-you tend to make your money in the early years, then move onto management before you are too old and too tired to continue with the foot in the door technique and the pattern. The same thing goes, to a certain extent, for advertising. <u>But other careers move at a different pace</u> become head-quarter in a museum, for instance, or head librarian, may take years.	many miss the right moment try to consider your position a year ahead

例6在文章的旁边做了边注。由于本文是一篇较短的文章，不可能将摘要应包括的内容全包括进去，但是其方法是通用的。做边注是为了将论文中的重点按文章结论的层次逐一挑选出来。例6的主题及主要内容均已边注，如果是一篇内容较长或很长的论文，则可对其次标题和二级小标题进行摘记，然后将有关重要内容充实进去。下面的例7是根据例6的原文内容及边注完成的一篇摘要。

【例7】

Abstract

In this article on successful careers, it says that it is important to know when to change jobs. Many people miss the right moment; so you should always think about where you are now, and where you will be in a year. Some jobs, though, move slowly, while others move quickly—careers have different time—scales.

从例7看，摘要的内容是主题与边注中的内容的结合。

（3）具体翔实。摘要的每个论点都要具体鲜明，直接讲论文"说明什么"，不要使用一些概念模糊的词汇。如不写为 not all the primary minerals are detrial，some are formed in place，因为其中的 some 使人感到费解，不知是指哪一些，造成了意义上的模糊，因此应在句中明确写出: not all the primary minerals are detrial，at least a past of the iron carbonate, titanate, and feldspars are formed in place.在后一句中将 some 的内容具体化了，具体到碳酸铁矿、钛酸铁矿和长石，这样，内容就具体了，也便于读者对论文进行检索。

（4）完整。完整实际上也是准确、清楚。摘要的内容要完整，这是因为读者是利用摘要或索引进行研究工作的，因此在摘要中一定要说明论文的主要内容，不要使用论文的导言或者某张插图来代替摘要中的内容。

5. 摘要的位置

摘要通常放在正文的前面，但是它的完成是在论文写作之后，而不是论文写作之前。这是因为论文在写作过程中要经过反复修改，内容在修改中有可能进行调整。所以，在论文写完后再写摘要是比较恰当的。

以下是两篇论文的摘要,供读者在写作实践中参考。

【例8】

The article investigates the factors affecting Chinese postgraduates' English oral proficiency. Little practice and fear of errors and unscientific classroom management are found to be the major factors. Also an emphasis on written English instructions seems to be another root cause. A series of methods such as how to organize classroom activities, how to allow the errors in students' oral practice, how to meet the needs of the learners and how to improve their oral abilities are presented.

【例9】

This paper discusses the problem of how to estimate a common mean parametric matrix relative to matrix loss. It gives six different optimality criteria and proves them to be identical in the class of homogenous linear estimations and gets a necessary and sufficient condition for their common admissible estimation. However, in the class of homogenous linear estimates, it will not be true any longer. But those criteria can be divided into two identical subclasses and a necessary and sufficient condition for their common admissible estimation is respectively given to each subclass.

二、关键词

关键词(key words)也称主题词,它是为了检索的需要,从论文中选出的最能代表论文中心内容特征的词或词组。这主要是为了检索刊物编制索引和输入计算机检索系统,方便刊登该文的刊物编制年终索引;同时也有助于读者了解该论文的主题及编排个人检索卡片。

按照惯例,摘要的末尾要有关键词,表明文章的特性,其位置是在摘要下方空1~2行。关键词选择的标准有两个,一是根据最新的权威性专业词汇(如中国科学院科学出版社出版的专业词汇)从论文中选出;二是根据"自己词汇表"从论文中选出。关键词一般为3~10个。

关键词是论文信息最高程度的概括,是论文主旨的概括体现。因此,选择关键词必须准确恰当,必须真正反映论文的主旨。关键词选择不当就会影响读者对论文的理解,也会影响检索效果。选择关键词的方法如下:

(1)要认真分析论文的主旨,选出与主旨一致,能概括主旨,使读者能大致判断论文研究内容的词或词组。

(2)选词要精练,同义词、近义词不要并列为关键词,复杂的有机化合物一般用基本结构的名称做关键词,化学分子式不能用作关键词,数学公式也不能用作关键词。

(3)关键词的用语必须统一规范,要准确体现不同学科的名称和术语。

(4)关键词的选择大多从标题和摘要中产生。

但是,如果有的标题和摘要没有提供足以反映论文主旨的关键词,还需从论文中选择。一些新的、尚未被出版社出版的专业词汇所收录的重要术语,也应选作关键词。

例如,在一篇关于形状记忆合金螺旋弹簧设计方法示例分析的论文中,其关键词就应选

Shape Memory Alloy（形状记忆合金）；Coil Spring（螺旋弹簧）；Designing Methods（设计方法）；Example Analysis（示例分析）。

在一篇关于非线性回归模型的应用的论文中,其关键词就应选做 Nonlinear Regression（非线性回归）；Models（模型）；Estimation of Parameters（参数估计）；Prognosis（预测）。

关键词的写法是每个关键词与下一个关键词之间用分号分开,最后一个关键词没有标点符号,每个词或词组中的任何一个实词的首字母一般要大写。

【例 10】
Copper；Cupric Sulfate；Production Method

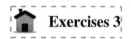

Please write down an abstract and key words of your article.

Overview of the Environmental Impact Assessment Process

What are the stages of Environmental Impact Assessment?

There are 5 broad stages to the process:

Screening

Determining whether a proposed project falls within the remit of the regulations, and whether it is likely to have a significant effect on the environment and therefore requires an assessment.

Scoping

Determining the extent of issues to be considered in the assessment and reported in the Environmental Statement. The applicant can ask the local planning authority for its opinion on what information needs to be included (which is called a "scoping opinion").

Preparing an Environmental Statement

Where it is decided that an assessment is required, the applicant must prepare and submit an Environmental Statement. The Environmental Statement must include at least the information reasonably required to assess the likely significant environmental effects of the development listed in regulation 18(3) and comply with regulation 18(4).

To help the applicant, public authorities must make available any relevant environmental information in their possession.

To ensure the completeness and quality of the Environmental Statement, the developer must ensure that it is prepared by competent experts. The Environmental Statement must be accompanied by a statement from the developer outlining the relevant expertise or qualifications of such experts.

Making a planning application and consultation

The Environmental Statement (and the application for development to which it relates) must be publicized electronically and by public notice. The statutory "consultation bodies" and the public must be given an opportunity to give their views about the proposed development and the Environmental Statement.

Decision making

The Environmental Statement, together with any other information which is relevant to the decision, and any comments and representations made on it, must be taken into account by the local planning authority and/or the Secretary of State in deciding whether or not to grant consent for the development. The public must be informed of the decision and the main reasons for it both electronically and by public notice.

（*https://www.gov.uk/guidance/environmental-impact-assessment#overview-of-the-environmental-impact-assessment-process*）

Unit 6 Environmental Laws, Regulations and Sustainable Development

Lesson 1　Governance — Water on the move

Part 1　Reading

From its source in the Black Forest in Germany to its delta on the Black Sea coast, the Danube crosses mountains, valleys, plains, countless towns, including Vienna, Bratislava, Budapest and Belgrade, and 10 countries. In its journey of almost 3 000 kilometres, the Danube converges with tributaries carrying water from nine additional countries[①]. Today, millions of people across the European continent are connected in one way or another to the Danube and its tributaries.

What happens upstream has an impact downstream, but not only. It is clear that pollutants released upstream will be transported downstream, but ships travelling upstream can facilitate the spread of alien species, such as the Asian clam moving westwards in the Danube, which can colonise large areas often at the expense of native species. When pollutants or alien species enter that water body, they instantly become a shared problem.

1　Governance Beyond the Land Mass

Current governance structures are almost entirely based on a common allocation of the land mass into territories. We can agree on common rules that apply within a defined area and set up bodies to enforce these common rules. We can even agree on economic zones at sea and make claims to the resources those areas contain. Certain vessels can be authorised to fish in those zones; companies can be granted rights to explore minerals in the seabed. But what happens when the fish migrate north or floating islands of plastic wash up on your shores?

Unlike the land mass, water is in constant motion, whatever its form may be, from a single raindrop to a strong ocean current or storm surge. Fish stocks and pollutants, including invisible chemicals such as pesticides and visible pollutants such as plastics, do not respect geopolitical borders and economic zones defined by international agreements between states. Like the air we breathe, cleaner and healthier rivers, lakes and oceans require a wider approach to governance based on regional and international cooperation.

2　River Basin Management

The approach for wider cooperation is one of the key principles behind the EU's water policies. The EU Water Framework Directive [②] — one of the cornerstones of EU water legislation — sees a river system as a single geographical and hydrological unit, irrespective of administrative and political boundaries. The Directive requires Member States to develop management plans by river basin. Given that many of Europe's rivers cross national boundaries, these river basin management plans are developed and implemented in cooperation with other countries, including European countries that are not members of the EU.

The cooperation around the Danube is one of the oldest initiatives of transboundary water management, dating back to the late 1800s. Over time, the focus has shifted from navigation to environmental issues such as pollution and water quality. Today, the initiatives to ensure the sustainable use and management of the Danube are coordinated around the International Commission for the Protection of the Danube River (ICPDR)[③], which brings together 14 cooperating states (EU and non-EU alike) and the EU itself, with a mandate over the whole Danube river basin, which includes its tributaries as well as groundwater resources. The ICPDR is recognized as the body responsible for developing and implementing the river basin management plan for the Danube. There are similar governance bodies for other international river basins in the EU, including the Rhine and the Meuse.

The Water Framework Directive also requires public authorities to involve the public in decision-making processes in connection with the development and implementation of river basin management plans. Member States or river basin management authorities can carry out this public participation requirement in various ways. For example, the ICPDR carries out public participation mainly by actively involving stakeholder organizations and consulting the public during the development phase of river basin management plans.

Given their vast dimensions, governance of the oceans remains an even more complex challenge.

3　Oceans — From Trade Routes to Deep-sea Mining Rights

For most of human history, seas and oceans were a mystery to be explored by all seafarers. Traders, invaders and explorers used them as transport corridors, connecting one harbour to another. Controlling key harbours and the sea routes connecting them resulted in political and economic power. It was not until the beginning of the 17th century, at the height of national monopolies over certain trade routes, that this approach of exclusive access was challenged.

Dutch philosopher and jurist Hugo Grotius claimed in *Mare liberum* (*Freedom of the seas*) in 1609 that seas were international territory and no state could claim sovereignty over them[4]. Grotius's book has not only offered legitimacy to other seafaring nations taking part in global trade but also played a fundamental role in shaping the modern law of the sea. Until the early 1900s, a nation's rights covered the waters within a cannon shot (corresponding to approximately 3 nautical miles or 5.6 kilometres) of its coastline.

The international discussion that started over nations' right to access to sea trade routes has over time changed to a discussion over the right to extract resources. During the 20th century, almost all countries[5] extended their claims. These claims vary between 12 nautical miles (22 kilometres) of territorial waters to 200 nautical miles (370 kilometres) for exclusive economic zones and 350 nautical miles (650 kilometres) for the continental shelf. The current international law is largely shaped by the United Nations Convention on the Law of the Sea (UNCLOS)[6], which entered into force in 1994.

In addition to introducing common rules for defining different national jurisdiction zones, the Convention stipulates that states have the obligation to protect and preserve the marine environment and calls for international and regional cooperation. Moreover, the Convention refers to the principle of common heritage of mankind, which holds that cultural and natural heritage in defined areas (in this case the sea bed, ocean floor and subsoil) should be preserved for future generations and protected from exploitation.

In such complex governance structures, it is always a challenge to agree on common rules and strike the right balance between protection of the natural heritage and economic interests.

The Convention's ratification took almost two decades, mainly due to disagreements over ownership and exploitation of minerals in the deep sea bed and ocean floor. The Convention established an international body, the International Seabed Authority, to control and authorise mining exploration and exploitation in the sea bed beyond the limits of the area claimed by countries.

Other governance structures and conventions cover different aspects of ocean governance. For example, the International Maritime Organization (IMO)[7] is a United Nations agency specialising in shipping, and it works, among other things, on preventing marine pollution caused by ships. Initially, its marine protection work focused mainly on oil pollution, but in recent decades it has been extended through a number of international conventions to cover chemical and other forms of pollution, as well as invasive species transported by ballast waters.

Pollution in water can be due to pollutants released directly to water or released to air. Some of those pollutants released into the atmosphere can later end up landing on land and water surfaces. Some of these pollutants affecting aquatic environments are also regulated by international agreements, such as the Stockholm Convention on persistent organic pollutants, the Minimata Convention on Mercury and the Convention on Long-Range Transboundary Air Pollution[8].

4 Governance in Europe's Seas — Global, European and Regional

The EEA report State of Europe's Seas concludes that Europe's seas can be considered productive, but they cannot be considered "healthy" or "clean". Despite some improvements, some economic activities at sea (e.g. overfishing of some commercial fish stocks and pollution from ships or mining) and pollution from land-based activities are increasingly putting pressure on Europe's seas. Climate change is also adding to these pressures.

Some of these pressures are linked to activities carried out outside the EU's borders. The reverse is also true. Economic activities and pollution originating in the EU has impacts outside the EU's borders and seas. Regional and international cooperation is the only way these pressures can be tackled effectively.

In this context, it is not surprising that the European Union is a party to the UN Convention on the Law of the Sea. In such cases, EU laws conform to international agreements but set specific objectives and governance structures to manage and protect common resources. For example, the EU Marine Strategy Framework Directive aims to achieve good environmental status in Europe's seas and protect the resources upon which economic and social activities depend. To this end, it sets overall objectives and requires EU Member States to develop a strategy and implement relevant measures. The common fisheries policy sets common rules for managing the EU's fishing fleet and preserving fish stocks.

Similar to international agreements, the EU's marine policies call for regional and international cooperation. In all of the four regional seas around the EU (the Baltic Sea, the North-East Atlantic, the Mediterranean Sea and the Black Sea), EU Member States share marine waters with other neighbouring coastal states. Each of these regional seas has a cooperation structure set up by different regional agreements.

The EU is a party to three of the four European regional sea conventions: the Helsinki Convention for the Baltic Sea; the OSPAR Commission for the North-East Atlantic; and the Barcelona Convention for the Mediterranean Sea. The Bucharest Convention for the Black Sea needs to be amended to allow the EU to accede to it as a party. Despite their varying ambition levels and slightly different governance structures, all these regional sea conventions aim to protect the marine environment in their respective areas and to foster closer cooperation for coastal states and signatories.

At the global level, the UN Environment's Regional Seas Programme promotes a shared "common seas" approach among the 18 regional sea conventions around the world. The United Nations' 2030 Agenda for Sustainable Development[9] also includes a specific goal, Sustainable Development Goal 14, Life below water, aimed at protecting marine and coastal ecosystems. The EU has been an active contributor to the 2030 Agenda process and has already taken measures to start its implementation.

5　When Stakes Go Beyond States

Common objectives and rules work best when implemented properly and respected by all those involved. National authorities can set fishing quotas but their implementation relies on fishing fleets. Using illegal gear, taking fish smaller than the minimum size allowed, fishing in other countries' waters or overfishing cannot be eliminated without compliance by fishermen and enforcement by authorities. The impacts — in this case, a decline in fish populations, a rise in unemployment in fishing communities or higher prices — are often felt by larger parts of society and across several countries.

Recognizing that various stakeholders impact the overall health of oceans, discussions previously led by governments have increasingly been involving non-state stakeholders. At the latest United Nations Oceans Conference held in June 2017 in New York, governments, non-state stakeholders, such as academia, the scientific community and the private sector, made close to 1 400 voluntary commitments to take concrete action to protect the oceans, contributing to Sustainable Development Goal 14. One of these commitments was made by nine of the world's largest fishing companies, with a combined revenue of about one third of the top 100 companies in the fishing sector. They pledged to eliminate illegal catches（including the use of illegal gear and catches over quota）from their supply chains. As more companies and people make such pledges and take action, together we could make a difference.

（https://www.eea.europa.eu/signals/signals-2018-content-list/articles/governance-2014-water-on-the-move）

🏠 Words and Expressions

governance	[ˈɡʌvərnəns]	n. 管理，治理
tributary	[ˈtrɪbjəterɪ]	n. （流入大河或湖泊的）支流
upstream	[ˌʌpˈstriːm]	adv. 向（或在）上游，逆流
Asian clam		n. 亚洲蛤
allocation	[ˌæləˈkeɪʃn]	n. 划拨的款项，分配的东西，划拨，分配
land mass		n. 陆地，陆块
authorise	[ˈɔːθəraɪz]	v. 批准，授权
irrespective	[ˌɪrɪˈspektɪv]	adj. 不考虑，无关的
mandate	[ˈmændeɪt]	n. （政府或组织等经选举而获得的）授权，（政府的）委托书，授权令；v. 强制执行，委托办理，授权
the Rhine		n. 莱茵河

the Meuse		n. 缪士河，默兹河
nautical miles		n. 海里
seafarer	['sɪfɜrɜ]	n. 海员
legitimacy	[lə'dʒɪtəməsɪ]	n. 合法性
ocean floor		n. 洋底
subsoil	['sʌbsɔɪl]	n. 底土
convention	[kən'venʃn]	n. 习俗，常规，惯例，（国家或首脑间的）公约、协定、协议
stipulate	['stɪpjuleɪt]	v. 规定，明确要求
ballast	['bæləst]	n.（船中保持平衡的）压舱物，（热气球的）镇重物
the Baltic Sea		n. 波罗的海
signatory	['sɪgnətɔːrɪ]	n.（协议的）签署者、签署方、签署国
academia	[ˌækə'dɪːmɪə]	n. 学术界
revenue	['revənuː]	n. 财政收入，税收收入，收益

Notes

① From its source in the Black Forest in Germany to its delta on the Black Sea coast, the Danube crosses mountains, valleys, plains, countless towns, including Vienna, Bratislava, Budapest and Belgrade, and 10 countries. In its journey of almost 3 000 kilometres, the Danube converges with tributaries carrying water from nine additional countries.

参考译文：多瑙河源起德国黑森林，流向黑海沿岸的三角洲，途中横跨山脉、山谷、平原和无数城镇，包括维也纳、布拉迪斯拉法、布达佩斯和贝尔格莱德，以及 10 个国家。在将近 3000 km 的行程中，多瑙河与另外 9 个国家的支流汇合。

② The EU Water Framework Directive 欧盟水框架指南

③ The International Commission for the Protection of the Danube River (ICPDR)多瑙河保护国际委员会

④ Dutch philosopher and jurist Hugo Grotius claimed in *Mare liberum* (*Freedom of the seas*) in 1609 that seas were international territory and no state could claim sovereignty over them.

参考译文：1609 年，荷兰哲学家和法学家雨果·格罗修斯在《海洋自由》一书中声称，海洋是国际领土，任何国家都不能对其主张主权。

⑤ During the 20th century, almost all countries extended their claims. 只有约旦和帕劳两个国家，有些地区仍然适用 3 海里规则。

⑥ the United Nations Convention on the Law of the Sea (UNCLOS)联合国海洋法公约（海洋法公约）

⑦ the International Maritime Organization (IMO) 国际海事组织（海事组织）

⑧ Some of these pollutants affecting aquatic environments are also regulated by international agreements, such as the *Stockholm Convention on persistent organic pollutants*, the *Minimata Convention on Mercury* and the *Convention on Long-Range Transboundary Air Pollution*.

参考译文：其中一些影响水环境的污染物也受到国际协定的管制，如《关于持久性有机污染物的斯德哥尔摩公约》《关于汞的最低限度公约》和《关于远距离越境空气污染的公约》。

⑨ The United Nations' 2030 Agenda for Sustainable Development 联合国 2030 年可持续发展议程

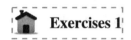

Exercises 1

1. According to the reading material, chose the best answer(s) from the options.

(1) According to the passage, the Danube crosses countless towns, including Vienna, Bratislava, Budapest and Belgrade, and _____ countries.

A. 3000　　　　　B. 12　　　　　C. 10　　　　　D. 4

(2) The EU Water Framework Directive sees a river system as a single _____ unit.

A. geographical　　B. hydrological　　C. administrative　　D. political

(3) The Water Framework Directive requires public authorities to involve the public in _____ processes in connection with the development and implementation of river basin management plans.

A. designing　　B. legislating　　C. exploratory　　D. decision-making

(4) The international discussion that started over nations' right to access to sea trade routes has over time changed to a discussion over _____ .

A. the protection of the environment　　B. the pollution of the ocean
C. the exploration of the ocean　　D. the right to extract resources

(5) Europe's seas can be considered productive, but they cannot be considered 'healthy' or "clean" because of _____ .

A. some improvements　　B. some economic activities at sea
C. pollution from land-based activities　　D. climate change

(6) The United Nations' 2030 Agenda for Sustainable Development⑨ also includes a specific goal, Sustainable Development Goal 14, Life below water, aimed at protecting _____ and _____ ecosystems.

A. underground　　B. river　　C. marine　　D. coastal

2. List the laws and regulations mentioned in the article in English and Chinese.

Part 2 Translation

英语翻译中从句的翻译（三）——定语从句

定语从句，顾名思义，是由关系代词或关系副词引导的从句，在句中做定语，一般是对某一名词或代词进行修饰和限定。通常来说，被修饰的词往往称为先行词，而定语从句位于先行词之后。定语从句有时也称为形容词从句（adjectival clause），在英语中使用频率极高，科技英语中尤为如此。

在英语中，如果单个词用作定语，一般放在所修饰的名词之前，而定语词组或从句一般放在所修饰的名词之后，有时也被其他句子成分隔开，句子往往结构复杂、形式多变；而在汉语中，定语成分一般位于中心词的前面，而且少用长而复杂的定语。

从结构形式来看，英语的定语从句有许多种，如直接由关系代词（或关系副词）引导，由"介词+关系代词"引导，由"名词+介词+关系代词"引导。从主句与从句的关系，即按它与先行词在逻辑含义上的紧密程度，分为限制性定语从句和非限制性定语从句两大类。张家民、李彦将根据关系代词和关系副词所修饰的先行词及其在从句中所起的作用，将定语从句分为六种：间位型定语从句、后位型定语从句、分离型定语从句、省略型定语从句、总括型定语从句和前位型定语从句。为了表达一些复杂的概念和严密的逻辑关系，科技文体往往使用一些成分复杂、结构层次繁多的定语从句。例如：

(1) Hydrogen, which is the lightest element, has only one electron.

译文：最轻的元素氢只有一个电子。（间位型定语从句）

(2) The volt is the unit which is used for measuring potential and potential difference.

译文：伏特是用来测量电位及电位差的单位。（后位型定语从句）

(3) The sun heats the earth, which makes it possible for plants to grow.

译文：太阳晒热大地，这使植物有生长的可能。（总括型定语从句）

认识定语从句对分析科技英语，深化科技英语翻译有建设性的现实意义。以下就将定语从句分为五大类加以探析，即限制性定语从句、非限制性定语从句、状语化定语从句、特殊定语从句和分隔定语从句。

一、限制性定语从句

一般来说，限制性定语从句和所修饰的名词或代词（称为先行词）之间的关系十分密切，其在意义上是不可或缺的修饰语，在结构上往往放在先行词之后，并且与先行词之间没有逗号格开，在带有限制性定语从句的句子中，定语从句是整个句子中不可缺少的部分，对先行词起到限制和修饰的作用它一旦缺席，主句的意思就会不完整、意义表达不明确，要靠从句加以补充说明。例如：

(1) Scientists say the planet is already on the road toward climate changes that could disrupt the global economy and endanger life.

译文：科学家们说，地球的气候已经开始变化，这种变化有可能破坏世界经济并危及人类的生命。（顺译译法）

(2) Scientists think the first loss of ozone reduces the amount of solar energy (which) the atmosphere can take in.

译文：科学家认为，臭氧层的首次变薄减少了大气层能够摄入的太阳能量。（逆序译法）

(3) Mathematical and data processing techniques that employ digital computers have entered more aspects of our lives than most vigorous proponents dream of ten years ago.

译文：数字计算机的数学技术与数据处理，在我们的生活中广泛应用，这是十年前大多数赞助者所未曾梦及的。（综合译法）

(4) Filtration is a simple process of passing the liquid through a sieve in which the holes are too small to allow the passage of the solid.

译文：过滤是使液体通过筛子的简单过程，因筛孔特别小，固体物不易通过。（转译为原因从句）

二、非限制性定语从句

与限制性定语从句不同的是，非限制性定语从句只是对先行词起补充说明的作用，它与先行词之间的关系往往比较疏远。也就是说，如果没有非限制性定语从句，主句的意思仍然清晰明了。一般来说，非限制性定语从句与主句间有逗号隔开，并由关系代词 which 和关系副词 when，where 引导。由于它和限制性定语从句有类似之处，翻译的时候也可以采用一些较为类似的翻译方法。例如：

(1) Mechanical energy can be changed into electrical energy, which in turn can be changed into mechanical energy.

译文：机械能可转变为电能，而电能又能转变为机械能。（顺序译法）

(2) Perhaps this is the "death ray", which we often read about in science fiction.

译文：也许这就是我们常在科幻小说中读到的那种"死光"。（逆序译法）

(3) A chrome plated surface, such as that with which we are familiar in automobile parts and plumbing fixtures, takes a high polish that is not easily tarnished or scratched.

译文：镀铬表面，例如我们在汽车零部件和管道装置上所经常看到的那种具有很高光亮度的表面，它不易变色也不易产生痕迹。（综合译法）

(4) Electrons also flow in a television, where they are made to hit the screen causing a flash of light.

译文：电子也流入电视显像管，撞击荧光屏，产生闪光。（分译译法）

三、状语化定语从句

状语化定语从句（包括限制性定语从句和非限制性定语从句）是一种边缘化定语从句。从句式形态上来看，属于定语从句，但实际上却与主句在逻辑上形成一种状语关系，充当说明原因、结果、目的、时间、条件或让步等的作用。

在翻译过程中，对于这类句子的处理方法多采用转译译法，也就是按照其内在逻辑关系和语义关系译成汉语各种相应的复句，比如转译为汉语因果关系、目的关系、时间关系、转折关系、条件关系、让步关系等。例如：

(1) We know that a cat, whose eyes can take in many more rays than our eyes, can see clearly in the night.

译文：我们知道，由于猫的眼睛能比人的眼睛吸收更多的光线，所以猫在黑夜也能看得很清楚。（转译为原因关系）

(2) Aluminum-copper alloy which when heat-treated has good strength at high temperature is used for making pistons and cylinder heads for automobiles.

译文：由于铜铝合金经过热处理之后具有良好的高温强度，所以用来制造汽车发动机的活塞与汽缸头。（转译为原因关系）

(3) To find the pressure we divide the force by the area on which it presses, which gives us the force per unit area.

译文：力除以它所作用的面积，得出单位面积上的力，就可以求出压强。（转译为结果的分句）

(4) Nowadays it is understood that a diet which contains nothing harmful may result in serious disease if certain important elements are missing.

译文：现在人们已明白，如果饮食中缺少了某些重要成分，即便其中不含有任何有害物质，也会引起严重疾病。（转译为条件关系）

(5) Electronic computers, which have many advantages, cannot carry out creative work and replace man.

译文：尽管电子计算机有许多优点，但是它们不能进行创造性工作，也不能代替人。（转译为让步关系）

(6) Very loud sounds produced by huge planes which lay low over the land can cause damage to houses.

译文：巨型飞机低空飞行时产生的巨大轰鸣，足以摧毁房屋。（转译为表示时间关系）

四、特殊定语从句

特殊定语从句也称为特种定语从句。所谓特殊定语从句，就是指修饰整个主句或主句部分内容的非限制性定语从句（通常由 which 或 as 引导），它代表整个主句或主句中的一部分在从句中充当成分。以下就来讨论以 which、as 引导的特殊定语从句。

1. 由 which 引导的特殊定语从句

这类定语从句之所以特殊，是因为由 which 这个关系代词引导的定语从句不同于前面提到的定语从句。它没有具体的先行词，而是说明主句的部分内容或整个内容，包括"介词+

which""介词+ which+名词""名词等+介词(多数情况下为 of)+ which",这种定语从句一般对主句的内容做进一步阐述说明或起承上启下的作用,其功能相当于一个并列分句。英译汉时,一般可采取分译法,将 which 译成"这""从而""因而"等。此外,由于 which 前面还有不同的介词,所以具体情况可因前面的介词不同而有所不同。例如:

(1) Liquid water changes to vapor, which is called evaporating.

译文:液态水变成蒸汽,这就叫蒸发。

(2) A very great amount of energy is packed in the flashes of a laser beam, which enables the laser to be a very powerful tool.

译文:在激光束的闪光中聚集了非常强大的能量,这使激光成为一种威力巨大的工具。

(3) Scientists are making great efforts to study the ways by which transformation of energy is carried out in plants and animals, which will help develop the science of bionics.

译文:科学家们正在大力研究植物和动物体内能量的转换方式,这将有助于仿生学的发展。

2. 由 as 引导的特殊定语从句

一般来说,由 as 和 which 引导的非限定性定语从句, as 和 which 可代表整个主句,相当于 and this 或 and that。这两个词的用法是不同的, as 的位置很灵活,而 which 则多在句中, as 代表前面的整个主句并在从句中做主语时,从句中的谓语必须是系动词;若为行为动词,则从句中的关系代词只能用 which。

在由 as 引导的定语从句中, as 通常指整个主句的内容或主句的部分内容,其位置十分灵活,不仅可以位于主句之后,还可以位于主句之前抑或是中间,在多数情况下,都有逗号将其隔开。翻译时,往往也会采取分译的翻译方法,把关系代词 as 译为"正如……那样""这""如""像"等。例如:

(1) Negative feedback, as is described here, is most widely applied in the automatic controls of various kinds of mechanism.

译文:负反馈,就跟这里所描述的一样,非常广泛地应用在各种机械的自动控制装置中。

(2) As the name implies, the World Bank was formed to provide sound long-term loans for reconstruction and development.

译文:顾名思义,世界银行的成立就是为复兴和发展提供可靠的长期贷款。

(3) The body has natural defenses against these or organisms, as will be explained in chapter 5.

译文:正如在第 5 章将要说明的那样,人体具有抵抗这些生物体的天然防御设施。

五、分隔定语从句

通常情况下,英语中的定语从句是紧密跟随在其先行词之后的。但有时出于表达的需要,也会被其他成分分开,形成割裂式的修饰成分,这在科技英语中尤为如此。对于这种定语从句，关键在于正确理解定语从句所修饰的对象并对其逻辑关系加以分析判断，做到对句子结构的理解了然于胸。翻译时，需要灵活加以处理，充分发挥译者的创造性。例如：

(1) These limitations do not apply to operations like boring which are performed with simple point cutter.

译文：这些限制不适于像镗削那样的加工方式，因为那些加工是使用单刀刃具进行的。

(2) In recent years ways have been developed by which air can be safely used over and over in space.

译文：近年来创造了一些能在太空中安全地反复利用空气的方法。

(3) A generator has been produced which changes mechanical energy into electric energy.

译文：能将机械能变为电能的发电机已生产出来了。

(4) We have made a number of creative advances in the theoretical research and applied sciences which are up to advanced world levels.

译文：我们在理论研究和应用科学方面，获得了不少具有世界先进水平的创造性成就。

以上我们具体讨论了各种定语从句的基本翻译方法，也从中总结了一些关于如何翻译定语从句的规律。除了以上的这些定语从句形式之外，还有一种比较复杂的定语，即定语从句套定语从句，甚至是套用多个定语从句，对于这种定语从句群，我们只有分清每层定语从句之间的逻辑关系，才能正确译出。例如：

(5) Radioactive atoms are very valuable in all sorts of ways, because we can use them to do things that are not possible with ordinary atoms that are not radioactive.

译文：放射性原子在各方面都很有价值，因为我们可以利用这些原子做普通非放射性原子做不到的事情。

(6) Using atomic energy, we shall be able to build space ships that will fly into the vast emptiness that separates the earth from the other planets.

译文：利用原子能，我们可以制成宇宙飞船，飞往将地球和其他行星隔开的广袤无垠的虚空。

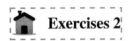

Exercises 2

Translate the following paragraph into Chinese.

(1) We can agree on common rules that apply within a defined area and set up bodies to enforce these common rules. We can even agree on economic zones at sea and make claims to the resources those areas contain. Certain vessels can be authorised to fish in those zones; companies can be granted rights to explore minerals in the seabed. But what happens when the fish migrate north or floating islands of plastic wash up on your shores?

(2) The international discussion that started over nations' right to access to sea trade routes has over time changed to a discussion over the right to extract resources. During the 20th century,

almost all countries extended their claims.

(3) Moreover, the Convention refers to the principle of common heritage of mankind, which holds that cultural and natural heritage in defined areas (in this case the sea bed, ocean floor and subsoil) should be preserved for future generations and protected from exploitation.

(4) One of these commitments was made by nine of the world's largest fishing companies, with a combined revenue of about one third of the top 100 companies in the fishing sector. They pledged to eliminate illegal catches (including the use of illegal gear and catches over quota) from their supply chains. As more companies and people make such pledges and take action, together we could make a difference.

Part 3　Writing

学术论文的撰写（十）——署名、目录、致谢、参考文献与附录的写作技巧

在科技英语写作中，对各个部分都有明确的要求和规定，都有其特殊的文体特点。下面就对署名、目录、致谢、参考文献和附录的写作要求和格式分别进行叙述。

一、署　名

论文要署名，这不仅是对作者的尊重和应有的荣誉，同时也表示文责自负，而且便于日后作为文献资料被索引和查阅。

论文的署名，根据美国《内科学纪事》有下述五个条件：

（1）必须参与过本项目研究的设计和开创工作，如在后期参加工作，必须赞同研究的设计。

（2）必须参加过论文中的某项观察和获取数据的工作。

（3）必须参加过对所观察的现象和所取得的数据的解释，并从中得出论文结论。

（4）必须参加过论文的撰写。

（5）必须阅读过论文的全文，并同意其发表。

若只对这项科研工作给予过支援和帮助，但不符合上述条件者不能列为作者。一般署名除作者姓名外，还要列出作者的专业技术职称、最高学历、工作单位、通信地址、邮政编码与电子邮箱，以便读者与作者联系。在几人同时署名的时候，署名的排列顺序一般按对论文的贡献大小排列。通常论文写作执笔者列首位，并在首页的脚注中注明。对于只按研究计划参加过部分具体工作、对工作缺乏全面了解的某一实验的参加者，只接受某项测试或常规分析工作的人，不应署名，但可在致谢中明确他们的责任和贡献。由几个单位协作完成的科研项目进行论文署名时，要同时写上各协作单位的名称。如果是某个单位的研究者以个人身份参与另一单位的研究工作，在论文中署名时，只需署被参与单位的名称即可。

学位论文作者的署名应先列出研究生的名字，然后再列出导师的姓名并注明导师的职称与头衔。凡是在国外期刊和国际学术会议上发表的论文要注明国家、省、市以及单位所在城市的名称。

署名写在文章题目的下方，在名字下面的圆括号内写上单位的名称。

署名一律使用汉语拼音，其正确的拼写方法如下：

（1）如果是三个字的名字，姓单独拼，名写在一起，如 WANG Zuoliang。

（2）如果是两个字的名字则将姓和名分开写，如：ZHOU Xing。

下面是一个采用多人署名的文章署名格式的实例，其目的是让读者能通过实例掌握正确的科技英语论文的署名方式，为在实践中结合具体情况进行运用提供一个标准格式

【例1】

Study on Biotransformation of Bioethanol from Lignocellulose by High Yield Cellulase-producing Aspergillus

LIU Yang[1], QIU Zhongping[2], FAN Chao[2]

（1. School of Geoscience and Environmental Engineering, Southwest Jiaotong University, Chengdu 610031, China; 2. School of Life Science and Engineering, Southwest Jiaotong University, Chengdu 610031, China）

在写作科技英语论文时，可根据例1的格式并结合实际情况进行取舍套用。

二、目　录

一般的论文没有必要对其内容列出目录(contents)，而有些较大课题的长篇综合论文(集)或学位论文，由于其篇幅较长，大多在标题后附有目录。目录反映的是论文的提纲。目录所列条目就是论文组成部分的小标题，并要逐项标明页码。当读者翻开目录时，就可以看出论文的大致轮廓，并能了解到各个论点之间的联系。目录也可帮助读者查阅章节，有的目录还列出参考文献、附录和索引所在的页次。

三、致　谢

由于科学研究工作常常不是一个人或几个人完成的，有时需要有关单位和个人的指导和支持。在论文后"致谢"中对给予过帮助的单位和个人表示感谢，以说明其在该项工作中的贡献和责任。

"致谢"通常由以下内容组成：

（1）对为研究工作提供方便和帮助的实验室或个人表示感谢，尤其是为研究提供专门实验设备或其他材料的人表示感谢。

（2）对经济上提供支持的人表示感谢。例如通过以赠送、签订合同或研究基金等形式为论文作者提供课题基金计划以外的资金援助的个人或机构，应对其表示感谢。

（3）对论文的内容提出过意见或建议，或者进行过其他形式的帮助，或者审阅过论文的人，应对他们的工作表示感谢。在致谢中应如实地说明他们所起的作用。

（4）对打字员、绘图员以及其他工作人员在论文中所做的工作，可以在致谢中提及。

【例2】

Acknowledgement

The author is indebted to Dr. Dale for suggesting the use of Ka kutani's theorem to simplify the proof and to the A. E. C for financial support.

【例3】

Acknowledgement

The author wishes to express his most sincere thanks to Prof. Gao Zhengheng, who read the manuscript carefully and gave valuable advice. Tremendous thanks are owed to Mr. Wu Xuehan, Wang Conghui and Jiang Yongcai for helping us with elemental analysis, mass spectral analysis and emission spectra recording. Also, the author is very grateful to Shanghai First and Twelfth Dyes Factories for offering some samples.

以上是两个表示致谢的实例。在两个实例中，作者均对为论文提供过帮助的个人或单位以及对他们所做的工作表示了谢意。

（1）表示致谢时常用的词汇与词组有：

be owed to; recognize; acknowledge; contribute to; help; contribution.

在表示致谢时开头常用的句型：

The author wishes to express his/her most sincere thanks to…

The author is indebted to…for…

A The author is very grateful to…for…

We would like to thank…for…

The author wishes to acknowledge…for…

We are gratefully to recognize…

（2）致谢的文体一般采用例2或例3的格式。但有时也采用先叙述被致谢的个人与单位在论文中所给予的帮助和所起到的作用，然后再表示感谢。使用后一种致谢文体时，往往是因为被致谢的个人与单位较多。

【例4】

Many at ERC have contributed to creating MPCS. In addition to the authors the original development group including Andy Baily, Dave Drake, Dave Cross Ethel Hopkins, Bill Jensen, Joe Kelly and Gordon Whitney.

Others, included Carol Feltz, Rod Gardner, Ken Larson and Mona Yousry developed test data collection and analysis software that link with MPCS. Some communication software was provided by Kurt Horton Winston Samaroo of ERC, along with Dave Hearon, Paul Baker and Warren Bailey from Network Systems laid out the high-level plan while John Basgall Wendy Clark, Danny Gallant, and Joan Wellman helped us bring the plan to life.

We gratefully recognize the contribution of these individuals and the many others who have since become part of the overall MPCS effort.

例 4 中，由于感谢的人较多，而且各项工作都有较多的人参与，只好分段将他们所做的工作分别列出，然后在致谢的最末一段对他们的工作表示感谢。

（3）在致谢中，一般使用过去时态来表示被致谢者在论文写作中或课题研究中所给予的帮助和支持。

四、参考文献

在科技论文之后列出参考文献是为了反映作者严肃的科学态度和所做研究工作的依据，也是维护知识产权的需要。所以，参考文献也是论文写作的一个部分。根据英语科技论文写作的规定，凡是引用其他作者的文章、观点或研究成果，都应标明，并在参考文献中说明出处。参考文献可以引用杂志论文、专利、毕业论文、专著等。引用要完整、清楚，以便读者需要阅读该文献时可以找到。参考文献的编写可以参考 GB/T 7714—2015，大多出版社对于参考文献也会有自己的要求，在编写这部分时，可以参考同一出版社的其他科技论文，也可以直接和编辑部联系以获得编写要求。

在列出参考文献时，首先要标出小标题 References。这个小标题既可放在中央，也可放在左对齐位置。

1. 表示参考文献的两种顺序

参考文献顺序的排列有两种形式。一种是按姓的首字母在字母表中的顺序进行排列；一种是按在正文中引用的顺序排列。

（1）按姓的首字母顺序排列

这种排列方式是一种传统的排列方式。在这种排列方式中，参考文献的格式与脚注的形式基本相同，不同的是将作者的姓放在第一个词的位置，并依照其字母的顺序来排列在参考文献中的顺序。对没有署名的参考文献可根据文章和书目的第一个词汇的字母顺序排列。如果是一个作者有两本著作或者论文，则按著作和论文的第一个单词的首字母顺序进行排列。

（2）按在正文中引用的顺序排列

按数字标号列出的参考文献在不使用脚注的情况下，可按数字顺序来排列参考文献。如果使用这种方法，所列出的参考文献应与文章中标注的顺序一致。这种排列方式与按姓的首字母排列顺序的方法一致，唯一不同的是姓名的书写方式没有顺序的变化。第二行只需同第一行文字的首字母对齐即可。

【例 5】

[1] Erdei B, Barta Z, Sipos B, et al. Ethanol production from mixtures of wheat straw and wheat meal[J]. Biotechnol Biofuels, 2010, 3(16): 1-9.

需要说明的是：①作者姓名中，姓用全称，名字这里只用第一个字母，但有的出版机构会要求写全名，如 James Robert Smith，引用格式：Smith J R；②一篇论文有三个以上的作者时，可只写前三个作者的姓名，其后加 et al.。

作者的姓名无论欧美均采用姓在前、名在后的著录形式。欧美作者采用没有重复的姓时，只著录作者的姓。有同姓不同名字的多个作者，不仅要著录其姓，还要著录其名字的首字母（大写），采用缩写时，省略符号点（.）。用汉语拼音书写人名时，姓全部大写，名字可缩写，取每个汉字的拼音首字母。

2. 引用不同种类的参考文献的表示方法

由于参考文献可包括杂志、著作、专利、毕业论文等，在参考文献中的格式也有所不同。下面我们以数字排序的方法对不同参考文献的表示方法分别进行叙述。

（1）引用期刊杂志论文。引用杂志论文的顺序是：①作者姓名；②论文标题；③[文献类型标识符/文献载体标识符].；④杂志名称；⑤出版年份；⑥卷号和期号（可省略掉 Vol 的字样）；⑦引用论文的页码。

【例 6】

Belal E B. Bioethanol production from rice straw residues[J]. Braz J Microbiol., 2013, 44(1): 225-234.

（2）引用专利。引用专利的格式是：①发明者姓名；②专利文献题名；③[P].；④国别与专利号；⑤专利生效日期。

【例 7】

Yamaguchi K, Hayashi A. Plant growth promotor and productionthereof[P]. Jpn, Jp1290606. 1999-11-22.

（3）引用毕业论文。引用毕业论文的格式：①作者姓名及论文名；②[D].；③论文保存城市；④保存地点；⑤论文完成的年代。

【例 8】

Smith J. A study of the impact of US monetary policy on the global economy [D]. New York: Columbia University, 2012.

（4）引用书籍。引用的书籍分为两种类型：一种是专著，另一种是合著。但是其格式基本相同，包括：①作者姓名；②书名；③ [M].；④版次(第一版不著录)；⑤出版地：出版者；⑥出版年；⑦起止页码。要注意的是，书名不可缩写。

【例 9】

Smith J. The birds of australia[M]. 3rd ed., Amsterdam: Elsevier, 2004: 116.

多人合作的著作只录前三人，其他的著者用 et al.省略。

文献类型与文献载体代码规定见表 6.1，以单字母标识。

表 6.1　文献类型与文献载体代码规定

参考文献类型	标识代码
普通图书等专著 Monograph	M
期刊文章 Journal	J
学位论文 Degree	D
会议录 Collection	C

续表

参考文献类型	标识代码
报纸 Newspaper	N
报告 Report	R
标准 Standard	S
专利 Patent	P
数据库 DataBase	DB

五、附 录

有些内容与论文主题关系密切，自身有较好的完整性，但若加入正文中会影响文章的连贯性，这些内容可以放在附录中。这些内容包括复杂的计算、公式的推导、校核用的数据和表格、编写的具体程序代码等。如果附录不止一个，可用 Appendix A 或Ⅰ，Appendix B 或Ⅱ等对其编号，如本书就有两个附录。

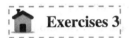

Please accomplish the article with the contents mentioned in Part 3 .

Outdoor Air Quality in Urban Areas

By European Environmental Agency

The 7th EAP (EU, 2013) aims to significantly improve outdoor air quality and move closer to World Health Organization (WHO) guidelines (WHO, 2006) by 2020. Air pollution is the number one environmental cause of death in the EU, responsible for more than 400 000 premature deaths per year (EEA, 2017a). According to WHO studies (WHO, 2013, 2014), exposure to particulate matter (PM) can cause or aggravate cardiovascular and lung diseases, heart attacks and arrhythmias, affect the central nervous system and the reproductive system and cause cancer. Exposure to high ozone (O_3) concentrations can cause breathing problems, trigger asthma, reduce lung function and cause lung diseases. Exposure to nitrogen dioxide (NO_2) increases symptoms of bronchitis in asthmatic children and reduces lung function growth. Health-related external costs range from EUR 330 billion to EUR 940 billion per year, depending on the valuation methodology, with evidence on the impacts of chronic ozone exposure adding around 5 % to this total (EC, 2013).

Policy targets and progress

A chief cornerstone of the EU environmental acquis in the field of air quality is the Air Quality Directive (EU, 2008). This directive sets a number of air quality standards not to be exceeded by a certain year and thereafter.

The communication on the "Clean Air Programme for Europe" (EC, 2013) sets the short-term objective of achieving full compliance with existing legislation by 2020 at the latest, as well as the long-term objective of seeing no exceedances of the WHO guideline levels for human health. The most troublesome pollutants in terms of harm to human health are particulate matter (PM), nitrogen dioxide (NO_2) and ground-level ozone (O_3) (EEA, 2016).

Outlook beyond 2020

In 2013, the European Commission proposed a Clean Air Policy Package for Europe (EC, 2013), which aims to achieve full compliance with existing air quality legislation by 2020 and to further improve Europe's air quality by 2030 and beyond. As a result of this package, the 2001 National Emission Ceilings Directive (EU, 2001) was reviewed. The new National Emission Ceilings Directive (EU, 2016) establishes national emission reduction commitments applicable from 2020 and stricter commitments from 2030 for sulphur dioxide, nitrogen oxides, non-methane volatile organic compounds, ammonia and PM2.5. In addition, and as part of the package, a new directive, the Medium Combustion Plant Directive, was approved in November 2015 (EU, 2015). This directive regulates sulphur dioxide, nitrogen oxides and dust emissions from the combustion of fuels in medium-sized combustion plants (with a rated thermal input of 1 and up to 50 megawatts).

These new commitments, together with the on-going implementation of air quality improvement measures at national, regional and local levels, are expected to improve air quality in Europe. However, the changes in meteorological conditions due to climate change are expected to increase O_3 concentrations as a result of expected increased emissions of both specific O_3 precursors and emissions from wildfires, with the latter likely to increase under periods of extensive drought (EEA, 2015).

Finally, it is expected that the age group composition of the EU population will continue to shift towards higher numbers of the elderly because of continuing increases in life expectancy (Eurostat, 2016). The overall potential air pollution-related health impact of this change remains uncertain.

(https://www.eea.europa.eu/airs/2017/environment-and-health/outdoor-air-quality-urban-areas)

UNIT 6 Environmental Laws, Regulations and Sustainable Development

Lesson 2 Global Megatrends

Part 1 Reading

1 Why are global megatrends important for Europe?

Europe is bound to the rest of the world through multiple systems, enabling two-way flows of materials, financial resources, innovations and ideas①. As a result, Europe's ecological and societal resilience will be significantly affected in coming decades by a variety of global megatrends — large-scale, high impact and often interdependent social, economic, political, environmental or technological changes.

Many global trends have significant consequences for Europe. For example, demographic, economic or geopolitical developments elsewhere can influence the availability and price of natural resources and energy in Europe. Increasing environmental pollution in other world regions likewise contributes to direct environmental and human harm in Europe. For instance, although European emissions of ozone precursor gases have declined significantly in recent decades, measured concentrations of ground-level ozone have not fallen at most ground monitoring stations②. There is evidence that this is partly due to the long-range transport of precursor gases from other parts of the world (EEA③, 2014a).

Conversely, Europe contributes to environmental pressures in other parts of the world. Greenhouse gas emissions in Europe contribute to climate change impacts elsewhere and potentially far into the future. Globalized supply chains mean that European consumption contributes to pressures on ecosystems and communities in other areas of the globe, for example through threats to global freshwater quality and quantity, and the degradation of habitats and landscapes (Tukker et al., 2014).

A global-to-European perspective is relevant for European environmental policymaking because Europe's systemic environmental challenges and response options are increasingly shaped by global drivers. Similarly to other advanced economies, Europe's relative size and influence on the global economy is expected to decline in coming decades. This changing global setting presents both challenges and opportunities.

The development of some global megatrends and related impacts over coming decades is becoming better understood. However, many uncertainties remain, associated with multiple drivers and change factors that unfold differently across world regions and over time. Global megatrends can also be perceived in contrasting ways by different societal groups and stakeholders. Continued global population growth, for example, can be seen as either a boost or a burden for economic development; urbanization can be perceived as a source of growing pressures on ecosystems, or as an opportunity for more resource-efficient lifestyles.

These uncertainties notwithstanding, it is clear that "business as usual" is no longer a viable development path for Europe. Current lifestyles in Europe and other developed regions put excessive pressures on the environment. Furthermore, as a growing global middle class increasingly adopts the resource-intensive consumption patterns of advanced economies, the total environmental burden is rapidly moving beyond globally sustainable limits (Rockström et al., 2009). This represents a growing threat to future advances in living standards and increasingly raises questions about the fairness of the wealthy imposing highly disproportionate burdens on the global ecosystem.

These trends underline the need for action to reconfigure systems of production and consumption so that they operate within planetary limits and thereby ensure the well-being of current and future generations. In Europe, as elsewhere, efforts to manage environmental pressures, economic development and human well-being need to overcome the short-termism currently dominating political and economic thinking and embrace long-term, integrated, global perspectives instead.

2 Implications of global megatrends for Europe's ability to meet its resource needs

The resources that societies rely on to meet their basic needs can be classified into four major categories: food, water, energy and materials (EEA, 2013b). In addition, ecosystems are essential to ensure the availability and quality of these resource categories, as well as providing a range of other ecosystem services that shape human health and well-being.

More direct pressures on European ecosystem resilience derive from urbanization, in particular from landscape fragmentation due to urban sprawl and expanding transport infrastructure [④]. Between 1990 and 2006, industrial areas and infrastructure in Europe expanded by 45%, residential areas grew by 23%, but population increased by only 6% (EEA, 2013b). Moreover, European terrestrial ecosystems are expected to be increasingly affected by drought, wildfires, floods, glacier melt, or species extinctions due to climatic changes expected in the decades ahead.

Marine ecosystems are projected to suffer from the combined effects from temperature increases, ocean acidification and sea level rise. Although significant reductions in European pollution levels have been achieved, severe problems persist, such as the eutrophication of aquatic ecosystems due to high nutrient concentrations originating from sources such as agriculture and urban waste water systems.

3 Responding to global megatrends—challenges and opportunities for Europe

As the examples in the previous section illustrate, the boundaries between developments in Europe and other parts of the world are growing more blurred. Europeans are increasingly affected by changes in distant regions — some very sudden, others unfolding over decades. Policy planning and strategic decision-making must, therefore, reflect the long-term and global contexts. That means finding ways to produce sufficient food, water, materials and energy to meet the needs of a growing global population, while maintaining ecosystem resilience and services.

Words and Expressions

megatrend	['megə,trend]	n. 大趋势，巨大潮流（指社会变化）
multiple	['mʌltɪpl]	n. 倍数； adj. 数量多的，多种多样的
innovation	[ˌɪnə'veɪʃn]	n. （新事物、思想或方法的）创造，创新，改革，新思想，新方法
resilience	[rɪ'zɪliəns]	n. 恢复力，弹力，适应力，还原能力
interdependent	[ˌɪntədɪ'pendənt]	adj. （各部分）相互依存的，相互依赖的
demographic	[ˌdemə'græfɪk]	adj. 人口统计学的，人口学的
geopolitical	[ˌdʒiːəʊpə'lɪtɪkl]	adj. 地缘政治学的
precursor	[prɪ'kɜːsə(r)]	n. 前身，先驱，先锋
global driver		n. 全球驱动因素
contrast	['kɒntrɑːst]	n. 明显的差异，对比，对照； v. 对比，对照，出现明显的差异，形成对比
notwithstanding	[ˌnɒtwɪθ'stændɪŋ]	adv. 尽管如此； prep. 尽管，虽然； conj. 尽管
reconfigure	[ˌriːkən'fɪgə(r)]	vt. 重新配置（计算机设备等），重新设定（程序等）

implication	[ˌɪmplɪˈkeɪʃn]	n. 可能的影响（或作用、结果），含意，暗指，（被）牵连、牵涉
derive	[dɪˈraɪv]	v. 得到，获得，（使）起源，（使）产生
fragmentation	[ˌfrægmenˈteɪʃn]	n. 破碎（化），碎片化
sprawl	[sprɔːl]	vi. 蔓延，杂乱无序地拓展；n.（城市）杂乱无序拓展的地区，随意扩展，蔓延
acidification	[əˌsɪdəfəˈkeɪʃən]	n. 酸化，成酸性，酸化作用
blur	[blɜː(r)]	v.（使）变得模糊不清，（使）视线模糊，（使）看不清，（使）难以区分；n.（移动的）模糊形状，模糊的记忆
unfold	[ʌnˈfəʊld]	v.（使）展开，打开，（使）逐渐展现
context	[ˈkɒntekst]	n. 语境，（事情发生的）背景

 Notes

① Europe is bound to the rest of the world through multiple systems, enabling two-way flows of materials, financial resources, innovations and ideas.

参考译文：欧洲通过多种体系与世界其他地区紧密相连，实现了物质、金融资源、创新和思想的双向流动。

② For instance, although European emissions of ozone precursor gases have declined significantly in recent decades, measured concentrations of ground-level ozone have not fallen at most ground monitoring stations.

参考译文：例如，尽管近几十年来欧洲臭氧前体气体的排放量大幅下降，但大多数地面监测站测得的地面臭氧浓度并未下降。

③ The European Environment Agency (EEA) 欧洲环境署

④ More direct pressures on European ecosystem resilience derive from urbanization, in particular from landscape fragmentation due to urban sprawl and expanding transport infrastructure.

参考译文：欧洲生态系统恢复力面临的更直接压力来自城市化，特别是城市蔓延和交通基础设施扩张导致的景观破碎。

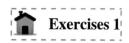 **Exercises 1**

1. According to the reading material, chose the best answer (s) from the options.

(1) Many global trends have significant consequences for Europe. For example, _____ ,

_____ or _____ developments elsewhere can influence the availability and price of natural resources and energy in Europe.

  A. demographic B. economic C. geopolitical D. independent

(2) Globalized supply chains mean that European consumption contributes to pressures on _____ and _____ in other areas of the globe

  A. individuals B. ecosystems C. administrators D. communities

(3) Current lifestyles in Europe and other developed regions put _____ pressures on the environment .

  A. excessive B. similar C. no D. less

(4) The resources that societies rely on to meet their basic needs can be classified into four major categories: _____ , _____ , _____ and _____ .

  A. food B. water C. energy D. materials

(5) More direct pressures on European ecosystem resilience derive from _____ .

  A. globalization B. acidification C. urbanization D. industrialization

(6) Policy planning and strategic decision-making must, therefore, reflect the _____ and _____ .

  A. short-term B. long-term C. local contexts D. global contexts

2. Explain the global megatrends to your friends in English.

Part 2 Translation

英语翻译中的转译

  转译是英汉科技翻译中经常采用的一种调整、变通甚至是创造性的翻译技巧。在翻译中，不仅词类可以转译，句子甚至是段落和篇章都可以转译，一个或多个词的词性发生了变化，其语法功能势必也会随之发生变化，那么，该词在译文中的成分和功能也会相应变化。比如，如果翻译中将动词转译为名词，那么修饰该动词的副词也会随之转译为形容词，转译后的名词可能成为主语，该句的句子成分就会随之发生重大变化。

  由于英汉两种语言不同的遣词造句和表达习惯，翻译中有时需要打破原文的句式结构，对译文的结构进行一番调整，使之符合汉语的表达习惯和规范。一般来说，句子成分的转换包括主语的转译、谓语的转译、宾语的转译、表语的转译、定语的转译、状语的转译、补足语的转译等方面。

一、主语的转译

(1) Both the frequency and inductance of a crystal unit are specified for inclusion in particular

filter.

译文：晶体单元要包含在一个特定的滤波器中，就要规定出频率和电感。（主语转译为宾语）

(2) The rotation of the earth on its own axis causes the change from day to night.

译文：地球绕轴自转，引起昼夜的变化。（主语转译为谓语）

(3) A semiconductor has a poor conductivity at room temperature, but it may become a good conductor at high temperature

译文：在室温下，半导体的电导率差，但在高温下它可能成为良导体。（主语转译为定语）

(4) Warm-blooded animals have a constant body temperature.

译文：温血动物的体温是恒定的。（主语转译为定语）

(5) May 1998 found him working in a research institute.

译文：1998年5月，他在一家研究所工作。（主语转译为状语）

二、谓语的转译

英语中的谓语在一般情况下不需要做特殊处理，但是，有时为了表达的需要，也可根据具体情况，将其转译为主语、定语和状语。例如：

(1) A highly developed physical science is characterized by an extensive use of mathematics.

译文：一门高度发展的自然科学的特点是广泛地应用数学。（谓语转译为主语）

(2) Today, London's tea market deal in tea from India, Sri Lanka and Africa.

译文：目前，伦敦茶叶市场经管的茶叶来自印度、斯里兰卡和非洲。（谓语转译为定语）

(3) For the first time in the annals of space, a piloted ship had succeeded in launching an earth satellite.

译文：载人飞船成功地发射了一颗人造地球卫星，这在航天史上尚属首次。（谓语转译为状语）

(4) In the past decade computer software has continued to advance very rapidly.

译文：在过去十年间，计算机软件一直在持续地高速发展。（谓语转译为状语）

(5) The power supply for the fire alarm system should be backed up by the emergency generator.

译文：火警系统的供电应有应急发电机作为备用电源。（谓语转译为宾语）

三、宾语的转译

英语中的宾语，在翻译时可以转译为汉语中的主语、谓语和定语等成分。例如：

(1) We need frequencies even higher than those we call very high frequency.

译文：我们所需的频率甚至比我们称之为甚高频的频率还高。（动词宾语转译为主语）

(2) There are three digital signal converters in our factory, each having its own features.

译文：我们厂有三台数字信号转换器，每台有其自己的特点。（介词宾语转译为主语）

(3) Physical changes do not result in formation of new substances, nor do they involve a change in composition.

译文：物理变化不生成新物质，也不改变物质的成分。（宾语转译为谓语）

(4) Much later Heinrich Hertz demonstrated radio waves in a primitive manner.

译文：隔了很久，赫兹用一种原始的方法证明了无线电波的存在。（宾语转译为定语）

(5) Another type of lens is thinner in the middle than at the edges, and is known as e concave lens.

译文：另一种造镜的中间比边缘薄一些，称为凹透镜。（介词宾语转译为主语）

四、表语的转译

由于英语中名词、代词、数词、副词、动名词和不定式等都可以做表语，而且英汉两种语言表达方式存在差异，表语的英译汉产生了不同的翻译方法。翻译时，往往需要转译为其他成分。例如：

(1) Rubber is a better dielectric, but a poorer insulator than air.

译文：橡胶的介电性比空气好，但绝缘性比空气差。（表语转译为主语）

(2) Television is different from radio in that it sends and receives pictures.

译文：电视和收音机的区别就在于电视发送和接收的是图像。（表语转译为主语）

(3) Computers are more flexible, and can do a greater variety.

译文：计算机的灵活性比较大，因此能做更多不同的工作。（表语转译为主语）

(4) This program was not popular with all of the troops.

译文：并不是所有军队的人都喜欢这个计划。（表语转译为谓语）

(5) A limited quantity of the reagents was available.

译文：可利用的药剂数量是有限的。（表语转译为定语）

(6) This problem is of great importance.

译文：这个问题非常重要。（表语转译为宾语）

五、定语的转译

英语中表示性质的定语，比如形容词或分词，往往暗含谓语的意味，翻译时，常常会被转译为谓语动词。此外，如果英语中某一名词转译成了汉语动词，那么修饰该名词的形容词或分词定语也要相应地转译为汉语的状语。例如：

(1) There are three states of matter—solid, liquid and gas.

译文：物质有三种状态：固态、液态和气态。（定语转译为主语）

(2) We should have a firm grasp of the fundamentals of a subject before making research into it.

译文：我们应首先牢固掌握某学科的基本知识，然后才能从事对它的研究。（定语转译

为状语）

(3) The earth was formed from the same kind of materials that makes up the sun.

译文：构成地球的物质与构成太阳的物质是相同的。（定语转译为表语）

(4) For quite a long time I thought that observation was merely another index of advancing age.

译文：在很长一段时间里，我觉得那种议论只不过表明另一迹象——我越来越老了。（定语转译为补语）

(5) Shortly after George Bush's election as President, his advisers were reported as recommending lower taxes and higher government spending.

译文：在乔治·布什当选总统后不久，他的顾问们就建议降低税收，扩大政府开支。（定语转译为谓语）

六、状语的转译

英语中有一些介词短语，常常在意义上和主语有密切的关系，表示主语的位置、状态和性质等，翻译时，往往将其转译为汉语中的主语。此外，如果被修饰词的词性发生变化，如英语的副词转译为汉语中的形容词，一些做状语的介词短语常常转译为定语。例如：

(1) Hydrogen-powered vehicles are also being developed in the United States, Canada, Japan and other nations.

译文：美国、加拿大、日本和其他一些国家也正在研制以氢为动力的车辆。（状语转译为主语）

(2) In size and appearance Mercury is very much like our moon.

译文：水星的大小和外观很像月亮。（状语转译为主语）

(3) In some countries the population declined, and governments actively encouraged people to have more children.

译文：有些国家的人口不断下降，因此政府积极鼓励人们多生孩子。（状语转译为定语）

(4) High-frequency waves that cannot be heard can be put to many valuable uses.

译文：人的听觉听不到的高频波也有许多有价值的用途。（状语转译为宾语）

(5) This satellite will go on orbiting for nearly twenty years.

译文：这颗人造卫星将继续沿轨道飞行20年左右。（状语转译为补语）

七、补足语的转译

英语中补足语分为主语补足语和宾语补足语，翻译时，往往将其转译为汉语中的谓语和宾语。例如：

(1) Water resources are reported to have great importance for hydroelectric power stations.

译文：据报道，水资源对于水电站具有重大意义。（补足语转译为谓语）

(2) They got the machine tool repaired in no time.

译文：他们很快把机床修好了。（补足语转译为谓语）

(3) Scientists often call this process dissolving.

译文：科学家们经常称这个过程为溶解。（补足语转译为谓语）

随着现代科学技术的发展，科技英语已经发展成为一种重要的英语文体，本书的 Part 2 简明扼要地阐述了翻译的基本理论知识，通过英汉两种语言的对比和大量的译例，主要在词汇、句法等方面介绍了科技英语翻译为汉语的一系列常用的翻译方法和技巧，为大家进一步探索科技英语翻译打下基础。同时，也希望这段学习能帮助大家更深刻地理解科技英语文献资料。如果大家能够结合本书的 Part 3 科技英语写作的部分一同学习，就有可能更恰当地表达自己，更充分全面地进行学术交流。

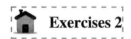

Exercises 2

Translate the following paragraph into Chinese.

(1) This raises questions about the limits of tolerable environmental pressure on the Earth's life support systems, sometimes referred to as planetary boundaries. Recognising that the total global environmental burden is indeed rapidly moving beyond globally sustainable limits, it is important to better understand how Europe's future ecological and societal resilience might by affected by current and future global trends, and conversely, how European systems of production and consumption are contributing to environmental pressures in other parts of the world.

(2) Increasing environmental pollution in other world regions likewise contributes to direct environmental and human harm in Europe. For instance, although European emissions of ozone precursor gases have declined significantly in recent decades, measured concentrations of ground-level ozone have not fallen at most ground monitoring stations.

(3) The successful delivery of these long-term visions and goals requires new and improved knowledge on the rapidly changing global context. This relates to both how Europe's environment and societies might be impacted by global megatrends and, conversely, how Europe's systems of production and consumption contribute to shaping global environmental pressures on a planet Earth with biophysical boundaries and limits.

(4) Most prominently, the Seventh Environmental Action Programme (7th EAP) — the strategic guiding document for all environmental policymaking in the current policy cycle until 2020 — sets out the vision of "Living well, within the limits of our planet", which directly relates to the idea of planetary boundaries.

Part 3　Writing

学术海报的制作

如果从事学术研究相关的工作，通常需要参加国际学术会议以便进行更广范围的学术交流。在国际学术会议中，一部分参会者会受邀做口头报告，这种机会较为珍贵。更多的参会者会受邀张贴学术海报或提交论文出版论文集，因此，把学术海报（Poster）做得规范、美观，对于提高论文的影响力非常有帮助。大型国际学术会议还会评审优秀 Poster，评选标准中除了研究内容本身的价值，Poster 制作的水平也很关键。

一、学术海报的结构

学术论文是制作学术海报的基础，在制作学术海报前还可以准备一份关于学术论文的 PPT。因为海报面积有限，能够表达的信息也有限，所以海报选择的一定是最重要的内容。组织信息时，除了文字以外，还要准备图片和表格。

学术海报的内容，总体上和论文大纲接近，通常有以下内容：

- **题目（Title）**
- **作者信息**
- 摘要（Abstract）
- **简介（Introduction/Background）**
- 问题（Problems）
- **目的（Objectives）或假设（Hypothesis）**
- **材料和方法（Materials and Methods）**
- **结果（Result）**
- 讨论（Discussion）
- **结论/总结（Conclusion/Summary）**
- 推论（Implication）
- 致谢（Acknowledgement）
- 未来展望（Future Works）
- **参考文献（References）**

具体制作时，可以根据论文、会议、参会人员等情况以及论坛允许的 Poster 尺寸做出取舍，但标记为粗体的部分通常是必须呈现的。

二、学术海报布局设计的基本原则

学术海报的具体要求，一般会由学术会议组委会提供。总体上来说，多数是横向布局（图 6.1），但如果是放到一个展示架子上，那么用纵向布局比较合适。

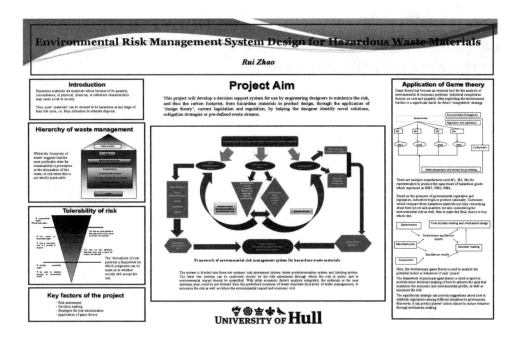

图 6.1　学术海报示例（一）

海报的具体尺寸很多样，标准的尺寸有 90 cm×120 cm、A0 尺寸或 36 cm×48 cm 等。确定了尺寸，建议先自行用一张 A4 白纸设计布局。如果经验不足，可以参考其他人的优秀作品，在网络上有很多优秀作品的展示，一些制作软件也提供成型的布局模板，可供选择。

1. 整体布局

（1）分区域，分模块。无论何种布局，一定要突出主题。用不同的颜色和边框等基本的元素，把确定要表达的内容区分为若干个模块。横向版面的话，可以分为 3~4 栏，纵向一般分为 2 栏；各栏之间一定要有间隔，明显区分，一般为 3 cm；海报的四周也要留白 4 cm 以上，避免打印不精确问题。

（2）内容不宜划分过多层次，从左到右、从上到下地表达。通常重点内容需要安排 1/3 以上的面积。

（3）大标题必须做到 3 m 外能够看清楚，如此才能吸引读者关注并方便其快速掌握论文的要点。

2. 颜色使用

（1）基本色的选择应该和论文主题以及领域的特点匹配，比如航空领域，选择一个蓝色系很合适，环境保护专业可以考虑绿色系。一般来说，学术海报的背景色会选用"纯白"无背景或明度、饱和度都较低的"暗"色，这样可以突出主题内容模块。无论背景色选择哪一种，主题内容模块需要和背景色有对比度，切忌选择相近的颜色。

（2）可以借鉴 PPT 软件提供的配色方案，方框、图表的颜色要与背景色协调，也可以使用渐变色增加视觉效果。

（3）使用的颜色不要超过 3 种（不包括图片）。

3. 字体使用

（1）正文字号至少 30 号，大段文字要按句分解成小段，便于读者阅读；标题要更大一些，为突出效果还可以选择与正文不同的颜色加以区分；主标题要最大，可选择 90~150 号。

（2）所有文字都要加粗。

（3）浅色背景上的深色字容易看清楚，如果有多层叠合，要注意是否影响同色系的文字的可见性。

4. 图表的使用

（1）尽量多使用图，图片最吸引人，能够传达的信息量也最大；图片所占面积通常超过 40%，文字内容要与其匹配协调（图 6.2）；图片质量要高，保证亮度及清晰度（200~300 dpi），避免拉扯失真。

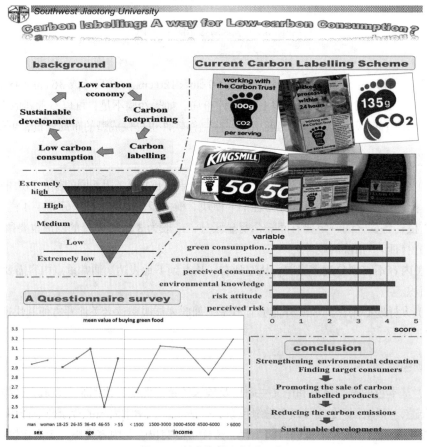

图 6.2　学术海报示例（二）

(2)图表一定要加上题注,也可以按顺序编号,避免造成读者的理解混乱。

5. 其他特别注意事项

(1)注重细节,设计完成后可先导出为 PDF 或者图片格式,反复检查错别字和其他文本错误,也可请别人帮助检查,发现自己的盲区。注意整个海报中的编号是否正确,文本框是否对齐,图片是否排列美观,表格里的数据是否用不同底色标出想要强调的部分。

(2)一定避免大段的纯粹文字,这样的海报鲜有吸引力。

(3)可以把论文的查阅地址标注在海报上,也可以使用二维码,方便有兴趣的读者进一步查阅全文。

(4)可以放上自己的照片并做简介,在照片下面或者海报最下面留下联系方式,方便有兴趣的阅读者与你联系。

三、制作打印

学术海报本质上就是一张宣传海报,所以专业的美术设计软件都适用于学术海报的制作,如 PS、AI 等。还有一些免费的或收费的专门制作学术海报的软件,可以提供给大家丰富的模板,大家可以自行搜索。但是,如果你未能掌握这些软件,也没有时间和兴趣单纯为制作一张海报去学习专业的美术设计软件,PPT 这个工具是完全足够的。当然,一些图片的处理可以用其他专业软件予以辅助。

打开 PPT→设计→页面设置,选择幻灯片方向,自定义幻灯片大小,更改高度、宽度,按照论坛要求或者自己的设计方案,选择一个尺寸即可(图 6.3)。

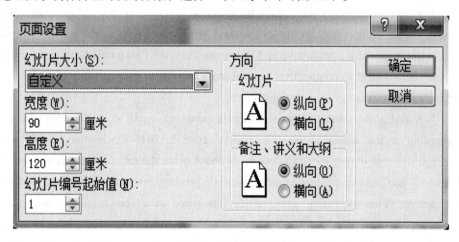

图 6.3 PPT 使用示例

然后就是具体的制作,可以参考 PPT 自带模板,也可以通过使用文本框,填充颜色,边框颜色等操作步骤,一步一步完成。打印时可以先把源文件转换为 PDF 或者 JPEG 格式,注意至少要 300 dpi 才可以保证打印效果。

 **Exercises 3**

Try to make an academic poster about your article or other issues you are interested in. Share your poster with your classmate and discuss on the whole process of study of academic writing. How will what you have learnt help you?

 Expanding Reading

Sustainable Materials Management Basics

What is Sustainable Materials Management?

Sustainable materials management (SMM) is a systemic approach to using and reusing materials more productively over their entire life cycles. It represents a change in how our society thinks about the use of natural resources and environmental protection. By examining how materials are used throughout their life cycle, an SMM approach seeks to:

- Use materials in the most productive way with an emphasis on using less.
- Reduce toxic chemicals and environmental impacts throughout the material life cycle.
- Assure we have sufficient resources to meet today's needs and those of the future.

How our society uses materials is fundamental to our economic and environmental future. Global competition for finite resources will intensify as world population and economies grow. More productive and less impactful use of materials helps our society remain economically competitive, contributes to our prosperity and protects the environment in a resource-constrained future.

U.S. and global consumption of materials increased rapidly during the last century. According to the Annex to the G7 Leaders' June 8, 2015 Declaration, global raw material use rose during the 20th century at about twice the rate of population growth. For every one percent increase in gross domestic product, raw material use has risen by 0.4 percent. This increasing consumption has come at a cost to the environment, including habitat destruction, biodiversity loss, overly stressed fisheries and desertification. Materials management is also associated with an estimated 42 percent of total U.S. greenhouse gas emissions. Failure to find more productive and sustainable ways to extract, use and manage materials, and change the relationship between material consumption and growth, has grave implications for our economy and society.

Sustainable Material Management's Life-cycle Perspective

By looking at a product's entire life cycle—from materials extraction to end-of-life management—we can find new opportunities to reduce environmental impacts, conserve resources, and reduce costs. For example, a product may be re-designed so it is manufactured using different, fewer, less toxic and more durable materials. It is designed so that at the end of its useful life it can be readily disassembled. The product's manufacturer maintains a relationship with its customers to ensure best use of the product, its maintenance and return at end-of-life. This helps the manufacturer identify changing needs of their customers, create customer loyalty, and reduce material supply risk. Further, the manufacturer has a similar relationship with its suppliers, which helps the manufacturer respond more quickly to changing demands, including reducing environmental impacts along the supply chain.

EPA's Sustainable Materials Management Program (SMM) Strategic Plan represents the collective thinking of EPA staff and management across the country, and includes stakeholder input from states, industry and nongovernmental organizations. The following three strategic priority areas chosen as the focus for EPA's future SMM efforts present significant opportunities to achieve environmental, economic and social results:

1. **The Built Environment** —conserve materials and develop community resiliency to climate change through improvements to construction, maintenance, and end-of-life management of our nation's roads, buildings, and infrastructure

2. **Sustainable Food Management** —focus on reducing food loss and waste and

3. **Sustainable Packaging** — increase the quantity and quality of materials recovered from municipal solid waste and develop critically important collection and processing infrastructure.

In addition to these strategic priorities, EPA will continue work in other SMM emphasis areas, including sustainable electronics management, measurement, life cycle assessment and international SMM efforts.

（*https://www.epa.gov/smm/sustainable-materials-management-basics*）

练习题参考答案

U1L1E1
1.（1）B （2）AC （3）D （4）A （5）D
U1L1E2
1.（1）C （2）C （3）ABD （4）B （5）AD
U2L1E1
1.（1）AB （2）A （3）C （4）ABD （5）B
U2L2E2
1.（1）ABCD （2）A （3）C （4）B （5）C
U3L1E1
1.（1）C （2）ABCD （3）ACD （4）B （5）D
U3L2E1
1.（1）AB （2）C （3）ABC （4）D （5）ABC （6）AC
U3L3E1
1.（1）CD （2）D （3）ABC （4）B （5）B （6）ACD
U4L1E1
1.（1）B （2）ABC （3）CD （4）ABCD （5）AD （6）B （7）ABCD
U4L2E1
1.（1）BCD （2）A （3）B
U4L3E1
1.（1）A （2）ABCD （3）BCD （4）ABCD （5）ABCD
U5L1E1
1.（1）AB （2）D （3）A （4）B （5）C （6）ABCD
U5L2E1
1.（1）ABCD （2）B （3）ABCD （4）ABCD
U6L1E1
1.（1）C （2）AB （3）D （4）D （5）ABCD （6）CD
U6L2E1
1.（1）ABC （2）BD （3）A （4）ABCD （5）C （6）BD

参考文献

[1] 王建武,李民权,曾小珊. 科技英语写作[M]. 3 版. 西安：西北工业大学出版社,2008.

[2] TAN Z C. Air pollution and greenhouse gases: from basic concepts to engineering applications for air emission control[M]. New York: Springer, 2014: 2-4.

[3] MURALIKRISHNA I V, MANICKAM V. Wastewater treatment technologies[M]//Environmental management. Amsterdam: Elsevier, 2017.

[4] 王春丽,官涤,米海蓉. 给排水科学与工程专业英语[M]. 哈尔滨：哈尔滨工程大学出版社,2016.

[5] 刘金龙,谷青松,刘晓民. 科技英语阅读与翻译[M]. 北京：国防工业出版社,2013.

[6] 游霞. 环境科学与工程专业英语[M]. 成都：西南交通大学出版社,2016.

[7] WU J. Environmental management in China — policies and institutions[M]. Beijing: Chemical Industry Press, 2020.

[8] 刘洋. 环境工程专业英语[M]. 成都：西南交通大学出版社,2008.